PENDULUM POWER

Reveals the mysterious origins of the pendulum, its history and uses and how a knowledge of its powers can help you communicate with the hidden levels of your being.

PENDULUM POWER

A mystery you can see, a power you can feel

by

GREG NIELSEN and JOSEPH POLANSKY

THE AQUARIAN PRESS
Wellingborough, Northamptonshire

First published in this format 1986

Original US edition published by
Destiny Books a division of
INNER TRADITIONS INTERNATIONAL LTD.
377 Park Avenue South, New York 10016, USA.

Copyright © 1977 by Greg Nielsen and Joseph Polansky

4 6 8 10 9 7 5

All rights reserved. No part of this book may be reproduced or utilized in any form or by any means, electronic or mechanical, including photocopying, recording or by any information storage and retrieval system, without permission in writing from the Publisher.

British Library Cataloguing in Publication Data

Nielsen, Greg
Pendulum power: a mystery you can see, a power you can feel.
1. Radiesthesia 2. Pendulum
I. Title II. Polansky, Joseph
133.3'23 BF1628.3

ISBN 0-85030-523-3

The Aquarian Press is part of the Thorsons Publishing Group

Printed and bound in Great Britain

Contents

Introduction

In the hand of a trained operator, the pendulum reads exact energy patterns, which in the final analysis, is the only truth we know.

No expert, authority, or sage knows *you* as well as you know yourself. When it comes to your inner world and the things that affect it, you are the authority. You are the expert. You are the god of your microcosmos, and you must take the responsibility for ordering it.

The pendulum is a tool for communicating with the deeper, more hidden levels of our being, the part of us which is, unfortunately, clouded by fear, ignorance, and false-to-fact opinions about ourselves and the universe we live in, the part of us that knows the truth because it *is* the truth. These levels of being are not conditioned by space and time and have powers which we as humans have not even begun to understand.

But this much is known. The more we link up with it, the more we identify with it, the more we allow its energy to flow through us and fill our minds with light, our hearts with love, and our hands with wisdom and power, the richer, simpler, and better our lives become.

The pendulum is simply one method of many for communing consciously with our deeper being and allowing it to guide our everyday actions. It is a tool of perception; that is, it gives us access to the extraordinary powers of perception of our deeper selves. These powers of perception are as superior to our ordinary five human senses as an H-bomb is superior to a firecracker.

The use of the radiesthetic sense, or pendulum power, is ancient. But most of mankind seems to have forgotten about it. In this book, we have attempted to provide a practical overview for beginners in this master science. We trace the historical thread of pendulum power from the dawn of prehistory onward. We show you how to make your own pendulums, along with step-by-step instructions on how to begin using it. And from our own personal experiences with the pendulum, we include a few examples we feel show some important practical applications of this science in everyday living. There is an extensive

Bibliography of the most important works in the field which serious students will find invaluable. This work is as comprehensive as we could make it in just one volume. No doubt when more people begin training with the pendulum and doing their own independent research, new facts and new applications will come to light.

Remember, we are dealing with a whole new sense, which means a whole new way of acquiring data from the environment – a new way of using the mind. The implications of this are so vast they amaze even the most fertile imagination. Pendulum Power, literally, can change the shape and structure of our entire civilization.

Nothing in this volume should be taken on faith. The pendulum is not a religion. Faith will come later after you have acquired some proficiency in tuning in, and it will be a healthier kind of faith. Right now adopt a scientific attitude. By this we mean a critical neutrality. Drop your prejudices and follow the instructions we have supplied. Take the necessary time and effort to practise and experiment with the pendulum in real life situations. Start with little things at first and as your confidence increases, use it in more important matters. Learn to keep your emotions calm and your mind neutral. All learning requires time. If you persist and practise even a few minutes a day for a year or year and a half (this is a variable depending on the person), you will be rewarded with a new sense, and because of this, you will become a more whole being. You will get more out of life because by your expenditure of time and toil you will have *become* more.

You are embarking on the first stages of an exciting voyage of self-discovery. Good luck.

1

What is the Pendulum?

What associations do most people have when they hear or see the word pendulum? Many immediately associate a pendulum with a clock, especially old grandfather clocks that tick-tock to the rhythm of a suspended pendulum. Others, more literary minded, think instantly of Edgar Alan Poe's weird and nerve-tingling tale *The Pit and the Pendulum.* In Poe's story the pendulum was a giant cutting device that gradually lowered closer and closer to a victim strapped to a table. Still others, perhaps more psychologically oriented, associate the word with the pendulum effect: the back and forth, up and down pattern that everyone's life seems to take. One day we are floating on cloud nine, the next we swing back – the pendulum effect – to a sullen or moody state of regret, resentment, or repression. Of course, there are those who, for one reason or another, have no associations with the word pendulum and who may want to check out the dictionary definition. Webster's *Seventh New Collegiate Dictionary* defines it as follows: 'A body suspended from a fixed point so as to swing freely to and fro under the action of gravity and commonly used to regulate the movements of clockwork and other machinery.'

Then, there are those who dream about pendulums. One person we know dreamed that he was a human pendulum turning clockwise over a group of desert chieftains seated at a banquet table. As he told us, 'I was checking them out.' How absurd, how fantastic, what nonsense may be your first reaction to our friend's dream. Nonetheless, in waking life he does use a device called a *pendulum* to 'check out' people, places, and things to determine their quality, that is, their vibration. A pendulum is a weight suspended from a string held between the thumb and index finger; and its movements indicate positive or negative energies.

Pendulum Power, the skill in handling a pendulum to measure both our inner and outer energies and force fields, is a modern outgrowth of the ancient art of divination or dowsing. It is by far, the most sensitive and accurate of all dowsing instruments, being equally handy in the laboratory, in the field, in the home, or on the job.

In the early part of this century L'Abbé Bouly, a priest of the sleepy French village of Haderlot, coined the word *radiesthesia* to describe the use of the pendulum. The name is a combination of the Latin *radius* for 'radiance' and the Greek *aisthesis* for 'sensitivity'. The word *dowsing* is often used interchangeably with radiesthesia but inappropriately so. Strictly speaking, dowsing refers to 'the search for water or minerals with a divining rod or pendulum', whereas *radiesthesia* covers both detecting and measuring the entire spectrum of radiations whether mineral, plant, animal, or human.

After World War I radiesthesia underwent a tremendous growth amounting to a full-fledged movement. Bouly, as well as another French abbé, Mermet, organized a series of congresses and conferences to increase the scientific knowledge of the many uses of pendulums, particularly the medical applications. Eventually, Mermet became known, in his lifetime, as the 'King of the Pendulumists', not only in France but also across the entire continent of Europe.

Men and women from around the world consulted Mermet: from France a poverty stricken widow whose son was missing, from Switzerland the chief engineer of a world famous corporation, from South America a concerned missionary, and incredibly, even the Vatican expressed genuine interest in Mermet's work, and requested guidance in solving archeological problems which baffled their experts. For his unfailing devotion and work the French National Society for the Encouragement of Public Welfare awarded him a prize.

In 1922, a pioneering American, Dr Albert Abrams, published a book about the application of the powers of the pendulum, which truly inaugurated the use of the pendulum in detecting and treating disease: the science of medical radiesthesia. In 1943, Dr Eric Perkins, Abrams' research assistant, delivered a lecture before the British Society of Dowsers describing Abrams' initial discoveries in psysiological radiations which opened his eyes to the possibilities of the pendulum's medical uses.

> Between six and seven P.M. a middle-aged, fairly healthy looking man came to consult Abrams about a sore place on his lower lip. He had had it for over two months; and it was growing larger and getting painful. Any physician would recognize the nature of the ulcer at a glance: it was an all too common form of cancer, usually spoken of as an epithelioma. Abrams told the patient to strip for a complete overhaul. Heart and lungs seemed normal enough. What

about the abdominal organs?

Abrams sat down in his chair in front of the standing patient, and began his examination by tapping – or as we say percussing – the patient's abdominal wall. I expect you have all seen doctors do this, but perhaps I had better explain. Supposing an area of the lung has become temporarily solid, full of some thick viscous secretion, as happens in pneumonia: when you tap over that area, you hear a flat, dull sound instead of the normal, resonant, full-of-air sound. Supposing there is a solid tumor growing inside the abdominal cavity: when you tap or percuss over it, you hear a flat, dull sound, whereas, normally, the note should sound hollow and drumlike because the abdomen is an air-filled cavity.

Now when Abrams tapped over an area just above this particular patient's navel, the note he elicited was just dead flat, so flat, so solid sounding as to suggest that a growth had developed behind the area which was being percussed. Cancerous patients are likely to develop what are known as secondary growths. Abrams told the patient to go and lie down on a couch so that he could carry his investigation a stage further by actually feeling for and defining with his finger tips the outline of any growth which might exist.

But no, rather surprisingly, there seemed no growth to feel. Abrams was not satisfied; he told the patient to get up and return to the place where he had first stood. Again Abrams sat down in front of the standing patient and tapped again, and still the note sounded so dull and flat as to suggest the presence of a solid tumor.

By this time, however, the setting sun was exactly framed in the window which the patient happened to be facing – the light was nearly behind him. Abrams noticed his patient's discomfort, and shifting his chair, Abrams said, 'turn half right and the glare from the sun won't be right in your eyes'. So the patient turned to face north, instead of west, and Abrams resumed his tapping. But behold, that consistantly dull-sounding note had been replaced by a note as ringing and as clear as if Abrams had been tapping on a drum. 'Somebody pull that blind down,' said Abrams, and then to the patient, 'face west again.' Back came the flat, dull-sounding percussion note. Over and over again the patient was asked to turn north while being percussed, then west again. While the patient was facing west, the percussion note sounded dull. When he faced south or north, the note Abrams elicited was invariably not dull, but resonant.

From this surprising experience Abrams began to realize that the human body is actually a kind of broadcasting station sending out messages – high-frequency radiations – from every cell, tissue, and organ. He learned that the pendulum could pick up, or tune into, these radiations and determine whether or not the vibration indicated health or disease. Unfortunately, as often happens with pioneers, Abrams was derided and scorned by most of his colleagues. Nevertheless, he forged ahead detecting disease, diagnosing, and treating his patients by picking up negative energies with a pendulum.

In France, new pendulum powers were discovered. André Bovis applied his divining talents to detect the freshness and quality of food. Also his research into radiations led him to agree wholeheartedly with Abrams on the body's radiation of positive and negative energies. Based on his relentless pendulum work, Bovis theorized that the earth has positive magnetic currents running north to south and negative magnetic currents running east to west. This would seem to have some connection with Abrams' discoveries with the patient who turned in different directions when being percussed. And as Bovis claimed, adamantly, subtle currents affected all structures on the surface of the earth. He found that *any* body placed in a north-south axis would become more or less polarized. His research showed that human bodies were curiously affected by these magnetic lines of force. The positive and negative energies, he concluded, entered through one leg and left by way of the opposite hand. Simultaneously, cosmic rays coming from outer space entered through the head and left through the other hand and foot. Also these vibratory currents, he discovered, emanated from the open eyes of each and everyone of us.

Bovis invented a specially sensitive pendulum with which to measure these mysterious rays. It was constructed from crystal with a fixed metal point suspended on a double thread of red and violet silk. The Frenchman from Nice referred to it as a *paradiamagnétique* because of its sensitivity to substances repelled or attracted by a magnet. Using his pendulum, he was able to determine the quality of foods by their vibratory radiations. Bovis put his skill to a practical use in his profession as a wine and cheese taster. Checking the varieties of wines and cheeses with the pendulum saved his tastebuds for the most delectable of cheeses and most fragrant of wines.

Bovis, Abrams, and Bouly were not the only pendulum prophets in the first half of the twentieth century. Paris, the city of love, art, and refinement was, and is, the Mecca, the holy city, where thousands of professed radiesthetists gather from every corner of the globe. Down a quaint and quiet side street, hidden from the tourists and Parisians alike, is an active old shop called the Maison de Radiesthesie. Alfred Lambert and his lovely wife have run the old shop for over five decades, stocking the shelves with every book, pamphlet, and leaflet available on the pendulum. Many of the finest volumes were written by distinguished French physicians. In fact, as one of our sources revealed to us, more than 2500 French physicians are actively and daily using the pendulum in their medical practices.

The shop also houses an incredible array of pendulums in every shape, substance, and size imaginable. Laying in drawers on velvet cushions are some of the world's most expensive and useful pendulums. Some are made of ivory taken from the tusks of African elephants. Some are shaped out of the most luxurious of jade, smuggled from Red China. Perhaps, the Lambert's best customer is M. Bourcart, one of the greatest exponents of the multiple uses of the pendulum.

Bourcart obtains extraordinarily accurate results, provided he uses the correct pendulum. His collection numbers over 1000! They are composed of a variety of materials, each one having been discovered by experiment to give the best results for him. The materials include such unlikely things as a monocle, a brazil nut, glass beads of various colours, and many other unusual objects.

Although Abbé Bouly's term for pendulum use, 'radiesthesia', has become the most popular word in the field, some researchers consider it inaccurate. S.W. Tromp, who wrote the book *Psychical Physics* in 1949, concludes from his extensive experiments that the existence of biological 'radiations' except for infrared are extremely improbable. He suggests that the term *radiesthesia* be replaced by 'electric fields', 'magnetic fields', or 'electro-magnetic fields'. He also recommends using the more neutral word *pallomancy*, meaning 'divination by the pendulum'. Others, including the Englishman Christopher Hills and the American mathematician Isidore Friedman, suggest a more structural name, 'radiational physics'. Since these other terms are more accurate and scientific, perhaps, in the future they will be more widely used. But at present, *radiesthesia* is the most popular term and is used by layman and scientist alike.

Bruce Copen, an Englishman who has been in the business of manufacturing and marketing pendulums since 1947, has made some staggering statements regarding the pendulum's uses. He firmly believes that about 90 per cent of the world's population could use the pendulum as a detector of radiations if they tried. Of the 90 per cent, he predicts that with training about half could become scientific researchers. However, to become an expert pendulumist requires having a keen intuition and a certain natural sensitivity.

According to Copen, radiesthesia or radiation physics, as your choice may be, can be used to enhance research in many scientific fields. At present, for example, geologists could use it in prospecting; farmers, in agricultural processes; and horticulturists, in cross-breeding. Some of the most valuable applications, he concludes, are research in medical diagnosis and treatment. It seems, in practice, that using the pendulum can cut time, energy, and money on any job to which it is applied. Furthermore, as Copen claims, 'In the right hands its accuracy is 100 per cent!'

Perhaps, to date, the most illustrious scientist to delve into the principles and practices of the pendulum was Nobel Laureate, Dr Alexis Carrel. More than forty-five years ago, while working under the auspices of the world-renowned Rockefeller Institute of New York, Dr Carrel realized the vital importance radiesthesia would have to the changing world. He stated his scientific opinion clearly: 'The physician must detect in every patient the characteristics of his individuality, his resistance to the cause of the disease, his sensitivity to pain, the state of all his organic functions, his past as well as his future. He must keep an open mind free from personal assumptions that certain unorthodox methods of investigation are useless. Therefore, he should remember that radiesthesia is worthy of serious consideration.'

In recent years, many methods, paths, and ways to self-development, self-realization, and self-transformation have come to light, and have been rediscovered enhancing millions of lives. The Chinese *I Ching*, astrology, meditation, and Jungian psychology have opened many creaky doors which otherwise may have remained sealed for hundreds of years. When put into practice in daily life, Pendulum Power can provide accurate and spontaneous information which can lead to happier, healthier, and more wholesome living. Whether in matters of business, education, love, art, or health, in the hands of a trained operator the pendulum can disclose the most precise course of action.

The ultimate consequences and benefits of the pendulum's power on the lives of its users cannot be calculated. It eliminates uncertainty in decision making and problem solving. Without a doubt Pendulum Power's practical value to anyone capable of using it in any field of endeavour can only increase the quality of life and work. Pendulum Power knows no boundaries – all who have the patience to perfect its use can benefit from it.

2

Jacob's Rod: The Mysterious Origins

The air was dry with the heat of the arid country of Canaan. The ground was sparse with grass as a dozen or so cattle sauntered about eating parched morsels. A man with strong shoulders and chest and intelligent eyes herded the scrawny cows, directing them with his stripped rod.

Stubbornly, they began moving at the goad of his mighty staff towards a green patch a half mile or so off to the north. The sun burned the rocky soil and dried the lips of the herder to cracking. The tough-hided cattle began picking up their pace moving as a herd towards the oasis. Their jaws were also dry and their sensitive nostrils sensed the sweet 'smell' of water.

Soon they reached the cool shade of the cypress trees and the moist pool of water at its centre. Then Jacob stepped to the pool's edge and dipped his stripped rod into the water. With his arm straight and his body emanating command, the cattle knew the water to be safe and began to drink.

From these ancient biblical times to the present, Jacob's Rod has evolved into the pendulum. The true origins are mysteriously veiled, but there are some extant records, stories, and tales etched on history's parchments.

In 1949, a robust band of French explorers and adventurers were searching the foothills of the Atlas Mountains. They stumbled across what are now known as the 'Caves of Tassili', caves with giant wall paintings. They found four prehistoric cave murals. The first depicted a group of men wearing Egyptian headgear and split skirts, who were rounding up a herd of cattle. The second showed a group of men crouching close to a fire, apparently preparing a meal. The third portrayed a ritual circumcision. The final painting was most remarkable since it depicted a dowser surrounded by his fellow tribesmen dowsing for water. These historic cave scenes have been dated by the carbon 14 process, proving them to be, beyond a doubt, at least 8000 years old!

As we have said, although dowsing, strictly speaking, is using a divining rod to locate underground water, minerals, and oil, still it is the ancient forebearer to modern radiesthesia. We can say dowsing is to radiesthesia what magic is to physics and what alchemy is to chemistry. It seems dowsing was known everywhere in ancient times. The Egyptians, Hebrews, Scythians, Persians, Medes, Etruscans, Druids, Greeks, Romans, Hindus, Chinese, Polynesians, Peruvians, and even the American Indian used some sort of rod for magical purposes.

In ages past, dowsing was referred to as *rhabdomancy* from the Greek *rhabdos* meaning 'rod' and *manteia* 'diviner or prophet', thus 'divination by rods or wands'. It meant the ritualistic practice of searching for springs, well sights, and precious metals buried in the earth.

There is very little mention of rhabdomancy by ancient scribes and historians because of its vital importance in the religious systems of the times. Temple priests kept their lips well sealed from the profane masses whom they knew would pervert the art and science of rhabdomancy.

From what can be deciphered and intuited from the hieroglyphs of Egypt the Pharaohs had great wisemen, sorcerers, and magicians in the temple who had rods which became serpents. This rod was called *Ur-Heka*, 'the great magical power'. Its shape was often that of a serpent without a head, or sometimes the handle had the head of a serpent united to the body of a ram.

In the Egyptian religious rituals, there is a hero who causes the water to gush forth. He says, in the character of the great one, who has been developed into a chief, 'I make the water to issue forth', or 'I make water to come'. According to Gerald Massey in his tomes *Ancient Egypt*, 'The striker of the rock with his rod or staff was Shu-Ankur, the impersonator of the force that burst up out of the rock at sunrise when the waters of day were once more set free.'

In Plutarch's *Life of Marcus Antonius*, from which Shakespeare gained most of his inspiration for the tragedy of Antony and Cleopatra, Cleopatra is described as essentially irresistible. She was a sensuous coquette, full of trickery and deceit. Her grief for Antony at the end was genuine enough, but blemished by her petulance, fear, and vacillation. She had characteristics of guile, selfishness, and probably greed. In fact, legend has it that the Egyptian queen had at least two dowsers with her at all times, whether at her palace or floating up the

Nile, not looking for water but for the treasure of treasures – Gold!

'And Moses was instructed in all the wisdom of the Egyptians; and he was mighty in his words and works.' (Acts 7:22.) That Moses was a skilled diviner and dowser is certain according to the second and fourth books of Moses, Exodus, and Numbers.

Exod. 4:17: 'And thou shalt take in thy hand this rod where with thou shalt to the signs.' Exod. 7:9 tells of Aaron's powers with the rod: 'When Pharaoh shall speak unto you, saying, show a wonder for you: then thou shalt say unto Aaron, take thy rod, and cast it down before Pharaoh, that it may become a serpent.'

The rod seems to have had, in the hands of its user, unlimited military powers. Exod. 17:9-11 tells of Moses's military leadership: 'And Moses said unto Joshua, choose us out men, and go out, fight with Amalek: tomorrow I will stand on top of the hill with the rod of God in my hand. So Joshua did as Moses had said to him, and fought with Amalek: and Moses, Aaron and Hur went up to the top of the hill. And it came to pass, when Moses held up his hand that Israel prevailed; and when he let down his hand, Amalek prevailed.'

In his great volumes, *The Rivers of Life*, J.G.R. Forlong refers to a certain wandering tribe called the Eduumeans. They, too, gained victory by holding up the 'wonder working rod of God'. The victory was not obtained by skill, number, or bravery but rather by the magical powers of the rod. In modern times, the pendulum has been used in military activities in Vietnam. Marines were trained to locate underground mines, ammunition dumps, tunnels, and enemy movements. Also there were some reports that during World War II, the pendulum was used by British Intelligence to determine Hitler's next offensive.

Moses's water wizardry is mentioned twice in the Old Testament. The first account is in Exod. 17:5-6: 'And Jehovah said unto Moses, pass on before the people, and take with thee of the elders of Israel; and thy rod, where with thou smotest the river, take in thy hand, and go. Behold, I will stand before thee there upon the rock in Horeb; and thou shalt smite the rock, and there shall come out of it, that the people may drink.'

The second account is from Num. 20:10-11: 'And Moses and Aaron gathered the assembly together before the rock, and he said unto them, hear now, ye rebels; shall we bring you forth water out of this rock? And Moses lifted up his hand, and smote the rock with his rod twice: and water came forth abundantly, and the congregation drank, and their cattle.'

Another authority, Rawlinson, in his third volume, *Ancient*

Monarchies, shows us that the rods of Aaron and Moses are the exact counterparts to the Egyptian rods, to the magic-working willow wand of the Skyths, and to the tamarisk rods of the Magi and present day Tartars. The rod of the ancients was a mighty symbol of the will, the ability to apply one's energy to a task until done. Today, the pendulum has become a tool of human knowledge turning to the desired information. The biblical heroes, it seems, had some power with the rod which was the historic source of present day radiesthesia.

The nomadic Arab tribes of old also dealt largely in divinational practices like *necromancy* and *rhabdomancy*. *Divination* can be defined as the art and science of consciously tuning to the lines of force projected into the future by peoples thoughts, feelings, and actions in the present. The conscious tuning in can be to higher beings, entities, and groups out of the human range of experience who know and understand something from a wider viewpoint. To the wandering Arab tribes, divination by rods, wands, almond sprigs, stocks, and staffs was an integral, sacred, and magical part of their lives. In fact, they carefully and exquisitely constructed special boxes and chests in which to carry them. Hosea of the Old Testament tells us of the stocks and staffs of Yahwah's 'holy place'. Perhaps this was some special artfully crafted chest, room, or closet that housed the rods.

Solomon, successor to the great King David, tried dowsing as a method for selecting the fairest of candidates for his harem. In those remote times, the temple prostitute was a woman well trained in the art of love and beauty. The king's seraglio had none of the onus that we today so automatically place on such a group of women. Often, the youthful beauties were chosen by the priest or magicians before they reached their teens. Then it took many years of rigorous and careful training in the arts of music, song, dance, cooking, and love making before they were worthy of the king. Perhaps, dowsing as practised by Solomon's wisemen to determine the quality of human love could open numerous doors to the modern-day bedroom.

During the time of Solomon, there were whispers and rumours throughout the land of a woman more beautiful than the fairest of the king's concubines. Ruler of her land, she was the Queen of Sheba. Even as Solomon heard of this woman, Sheba heard of Solomon as the wisest of men. Each, having diviners in their respective courts expert in the use of the rod, asked to know more of the other. The diviners verified the rumours.

Sheba set out to see Solomon taking with her camels bearing riches of rubies, exotic spices, and other precious metals too numerous to count.

Reaching the outskirts of Jerusalem, she made camp putting up hundreds of bright-coloured tents. The next day Sheba, carrying the treasures she brought as gifts, visited Solomon at his temple. Solomon was so struck by her radiance that he expounded for hours on the wisdom of his god, thereby lifting Sheba's soul to the heights of love.

When Sheba had returned to her camp. Solomon summoned the wisest and most accurate of his diviners. The diviner raised his rod and it trembled in his hand. The hour for their tryst would be that night just after midnight. It was the Lord's wish.

One Arabic manuscript dealing with this theme mentions: 'When the Queen of Sheba came to visit Solomon, she had, amongst her train, diviners who divined for gold and water.' Perhaps they, too, divined for the proper hour for their meeting in the queen's tent.

Another book in the library of an Englishman, Colonel Tillard, deals with the folklore of Upper Egypt. As the story goes, there was a mighty and rich king who had two daughters. One, named Dronker, was fair, most beautiful, truly a sight to behold. The other Asiut was dull, plain and downright ugly. When the king died he left his entire estate to Asiut, the ugly one. Dronker was left with a small area beyond the estate without water. The diviners heard of Donker's plight and seeing her radiance immediately found water. Such was the wisdom of the king.

The Middle East was by no means the sole source of dowsing and divining activity in ancient times. The Chinese, as far back as the reign of the Emperor Yu in 2200 B.C. used various divining methods. Marco Polo in his adventures and travels throughout the Orient discovered the use of some form of rod. It seems the Chinese were particularly adept at detecting what they called 'the claw of the dragon', that is, noxious or harmful rays.

Druids, too were extremely sensitive to vibrations, rays, magnetic forces; call it what you will. Just as Chinese diviners selected suitable sites for buildings, we can assume that Druid magicians with the power of the rod located the vibrationally correct site for building the giant structure, Stonehenge.

Independently, modern day radiesthetists like Abbé Mermet and V.D. Wethered have detected harmful radiations. In fact, it is common knowledge in most parts of the world that certain houses and streets are unhealthy. According the Mermet, the trouble is usually caused by the presence of unwholesome underground water. But as Wethered rightfully points out, there are other causes such as soil composition and air ionization intensity.

The Chinese were not the only Eastern culture to use the divining rod. A report of the rod being used to find hidden objects in the mid-1850s in India indicates its long-time use there: It seems that a Calcutta writer had some possessions stolen from his home. Having little success with the local authorities, he found a native famous for the divining of missing objects. The native proceeded with certain incantations which caused two other natives present to hold two bamboo rods parallel to each other. The writer is quoted as saying, 'We confessed to our great surprise, we beheld the bamboos crossing each other horizontally, and afterwards alternately rise up and descend.' Continuing this practice for a time, the direction in which the thief lived was determined. Later, the culprit was found.

A very similar practice has been discovered among the Maganja of Central Africa which probably dates back millenia. The particular episode described involved stolen corn. The medicineman chose four men to hold the rods, two men to each rod. One Anglo-Saxon's curiosity led him to ask the medicineman what it meant. The answer was: 'Wait and you will see. The rods will drag and drag the men, until they drag them to the person who stole the corn!' Eventually, they were led 'like mad things' to the hut of one of the wives of the robbed person.

Sir Frank Swettenham, describing the methods of divination used by the Maylays, writes: 'Yet another plan is to place in the hand of the pawang, magician, or medium a divining rod formed of three lengths of rattan, tied together at one end, and when he gets close to the person "wanted" or to the place where anything stolen is concealed, the rod vibrates in a remarkable manner.' How old this form of divination is we can only guess, but certainly it dates back at least two or three thousand years.

T.J. Hutchinson in his book *Two Years in Peru* tells of a carved figure in rock, bearing a forked rod. From archaeological reports it is estimated that Peruvian civilizations date back before 9000 B.C. No doubt the hidden divining power of the rod belonged to their sorcerers and magicians.

In a way it is peculiar that during Roman times there is no direct mention of water dowsing or divination by rods. Neither Vitruvius nor Pliny mention the rod even though they write at length about the means of discovering springs. However, one lost volume by the Roman Varro, *Virgula Divina*, probably dealt exclusively with methods of divination. Whether or not he discusses dowsing for water, lost objects, or gold is a matter of speculation. Again, Cassiodorus in the fourth century A.D. and Palladius in the sixth century praised waterfinders without specifically mentioning the use of the magic rod.

Not until the eleventh century is there another allusion to the rods. Notker, a monk in St Gallen and one of a long line of monks to delve into the mystery of the divining rod, writes of the volatile and mercurial rods. Later, in the fifteenth century Basilus Valentinus, a Benedictine monk devoted seven chapters of the second book of his work to a didactic account of the use of the divining rod.

From the twelfth to the fifteenth centuries, there are numerous Germanic mentionings of the rod but none specifically to the finding of water or minerals. Germanic tales, stories, and folklore refer to the golden rod of the Nibelungslied, the paradisial wishing rod of Gottfried of Strassburg, and to the magical rod or taming wand of the Edda.

One very interesting reference to the wishing rod is in Conrad of Megenberg's *Buch der Natur* written between 1348 and 1350. It seems when split hazel rods are used as spits, they turn by the heat of the fire alone!

Despite all these references to wands, staves, staffs, bamboo sticks, and the like dowsing with the rod, as we know it today, did not become an established art until the sixteenth century. However, there is one exception. A brief whisper of it in a manuscript dated 1430 written by a mine surveyor.

Before we unravel the written history of modern day dowsing, we want to mention a few of the remarkable water divining tales that circulated throughout Europe during the Middle Ages.

In Sweden, the dowser was called a *Dalkarl.* A Swedish twelfth-century manuscript says that the Dalkarl should find a mountain ash which has grown from a seed dropped out of a bird's beak. Then, at twilight, between the third day and night after Lady Day he should break off a twig. Next, he should bring the twig in contact with iron or steel making sure not to let it fall to the ground. At that point the rod is ready for various magical purposes.

In Denmark, they claim lost treasures can be discovered with a magic rod, called the *Finkelrut*, which is cut during St John's night, while invoking the Christian trinity. The Finkelrut may also be used for finding water but only if it is made of a willow branch and operated by a man born under Aquarius. Another fascinating note regarding the Finkelrut: in arms of the ancient Danish family of Bille, there is a figure of a troll holding a sapling. Tradition has it that during the dry season such a troll came with a sapling in his hands and found water.

There is a remarkable old folktale told long, long ago in a valley of the Carpathians. A woodcutter said to his wife, 'If we could but succeed in finding the wizard's rod, all our needs would have an end at one blow.' 'But how is to be recognized the tree from which the wizard's twig has to be broken?' asked the wife. 'For that,' replied the husband, 'you must walk through the whole wood, and at the hour of midnight you must listen to find which tree begins to sing; from that one you must then quickly break off a twig, and run as fast as you can to get out of the wood. Then, during the next full moon, you must run round the whole valley, striking every rock you pass with the wizard's rod. When you chance upon the right rock, it will split and form a cave. Going into it, you will find endless treasure, and then we are rich folk. All this was told to me by the mountain spirit, to whom I once rendered a great service. Often already I have spent nights in the woods, but always I have failed to find the tree.'

From this time both together often wandered through the woods. One night there was such a tempest outside that the wife did not venture out. Only the man huddled himself in his cloak and made his way out. Suddenly, he noticed that it was not raining where he stood, while barely two paces away the fearful tempest raged around him. Mechanically, he stayed in his place and awaited the departure of the storm. Thus approached the midnight hour. Then the tempest stopped, and as the woodcutter was about to go on, suddenly the tree which stood behind him softly began to sing. Quickly, he turned round, clambered up the tree, broke off a twig, and ran hurriedly away. He reached his home quite exhausted and showed his wife the find.

Scarcely had the next night of full moon come round, when he was already in the valley with his wizard rod, running around and knocking everywhere. One of the big rocks suddenly burst open. He was about to go in, but jumped back with fright as he saw a man standing before him. But the man nodded in a friendly way and so he went in, when he was almost blinded by heaps of diamonds. In his ignorance he took it all for glass and left it laying. Instead, he so loaded himself with gold that he could hardly drag it out, and threw down the rod. As he went out, the old man looked at him meaningfully to remind him that he had left the wizard rod behind. But the woodcutter thought he was being threatened, and ran away with his treasure. Only at home did he remember the little wand, and it was then clear to him why the old man had looked at him so meaningfully. Quickly, he ran back, but the rock was closed.

Tales from folklore are inexhaustible and could easily fill an entire book. We have chosen some of the more entertaining and amusing stories. There is an old saying to the effect that in every tale there is a bit of truth. It is difficult not to believe these stories were founded on actual events.

Figure 1. A Sixteenth-Century Dowser
G. Agricola, *De Re Metallica* (1556)

Returning to the reports of recorded history, in 1518 a conflict with the religious powers of Protestantism and Catholicism began which lasted 400 years. Martin Luther, one of the leaders of the Protestant Reformation in Germany, condemned dowsing as a form of 'black magic'. He actually issued a proclamation declaring the use of the rod as a breach of the First Commandment, 'Thou shalt have no other gods before Me.' Luther's condemnations appear odd when we think that his father, who was a miner, certainly must have been aware of the pendulum since it was in constant use in the mines at that time.

Many sources consider Germany, specifically the Harz Mountain district, as the birthplace of the modern divining rod. In Georgius

Figure 2. The Dowser Unmasked
T. Albinus, *Das entlarvete Idolum* (1704)

Agricola's *Latin-German Glossary*, in part of an essay published in 1530, there is a reference to the rod. However, it was not until 1556 that Agricola treated it more fully in his treatise *De Re Metallica* (See Figure 1).

During the reign of Queen Elizabeth (1558 to 1603) German miners were brought to England to develop the mining industry of Cornwall. Specifically, German dowsers were used to discover lost tin mines. By the end of the seventeenth century, the use of the divining rod for locating minerals and water had spread all over Europe and had aroused great controversy among scientists not to mention the clergy. The vast majority of dowsing opponents came by their attitude, not for scientific reasons but because they associated dowsing with satanic practices. Their beliefs, for the most part, were well founded for, in fact, most dowsers in the Middle Ages gave their practice an aura of mystery which really boiled down to greed. For example, one dowser's incantation went as follows: 'In the name of the Father and of the Son and of the Holy Ghost, I adjure thee, Augusta Carolina, that thou tell me, so pure and true as Mary the Virgin was, who bore our Lord Jesus Christ, how many fathoms is it from here to the ore?'

These rather questionable practices and misuses of the divining power make it understandable why the Church banned all forms of dowsing. The Protestants at Wittenbury in 1658 officially declared that the movements of the dowsing rod were caused by outright fraud or by some binding pact with the devil. In 1659, after the publication of a book by the Jesuit Father Gaspard, dowsing was hotly debated by churchmen. Some approved it, sighting great successes of pious monks; others threatened dowsers with excommunication, calling it the work of the devil (See Figure 2). In 1701, the Inquisition was able to cite divining misuse as a method of determining guilt. Thus, they decreed that dowsing for missing persons or criminals was to be considered sinful. Unfortunately, the more scientific and intelligent dowsers such as the Baron de Beausoleil suffered, as well. In 1642, he was charged with sorcery and imprisoned, where he died soon after.

On the first floor of the Science Museum at South Kensington, there is a collection of emblematic tools that were used by the miners' guilds in Saxony between 1664 and 1749. There are axe forms, with wooden hafts, and ivory and bone vignettes are cleft into the sides of the hafts. These vignettes are engraved with scenes of mining operations and other things. One vignette depicts a man with a dowsing rod and a second man with a dowsing rod facing a third man with a very large pendulum. This is the first hard evidence showing the use of a pendulum in divining except for a few undocumented references. In 1799, Professor Gerboin of the University of Strasbourg brought back a pendulum from India and presented it to the Academy of Sciences in Paris. He made a study of the pendulum over masses of metal, and recorded his findings in a book published at the beginning of the nineteenth century. Ritter, a German physicist, also published a book at about the same time, agreeing with Gerboin. In 1833, Ampère and Chevreul, famous French scientists of the time, were appointed to make an investigation. Unfortunately, it seems they were not sufficiently sensitive to the vibrations involved, thus making their conclusions negative. Their findings retarded the study of the pendulum for more than half a century.

Since that time, especially from the late nineteenth century to the present, numerous experiments, tests, and observations have been conducted using the scientific method. Not all the researchers lacked the keen sensitivity so necessary for divining rod work whether with the dowsing rod or pendulum. Their findings would fill endless volumes.

A great number of societies, organizations, groups, governments, and the like have devoted their attentions to the principles and practices of the divining rod; to mention only two, The British Society of Dowsers and The American Society of Dowsers. Among the many governments experimenting with dowsing are the Soviet Union, the United States, Canada, France, and England. In 1931, the government of British Columbia, Canada hired a dowser to locate water for the homesteaders of the area. The Bristol-Myers Company paid $2500 to a dowser for locating an underground water supply in New Jersey. Canadian industries, owned partly by the E.I. DuPont de Nemours Company also compensated a dowser $2500 for finding an adequate source of water. The RCA Company did not spare any expense paying a dowser for the same job at their Victor plant in New Jersey.

Even the most celebrated scientists of the twentieth century have looked into the phenomenon without prejudice, extreme scepticism or scoffing. Albert Einstein considered dowsing fascinating and believed that electro-magnetism would somehow give us some scientific answers to the why. A Duke University professor, Joseph B. Rhine, offered the explanation that dowsing is related to ESP more than to physics. Perhaps, in time, we will discover that it is more answerable by a combination of the two. Regardless, Charles Richet, a French Nobel Prize winner, summed it up by saying, 'Dowsing is a fact we must accept.'

3

Why the Pendulum Works

We live in an energy universe. And every organism is surrounded by all kinds of energy, some beneficial and some utterly destructive. To survive, every organism had to develop means by which it could sense these energies so that it could benefit from the former and avoid and protect itself from the latter. This ability to sense and distinguish different energies is a fundamental property of protoplasm. Plants will shrink away from people who radiate hostile energy towards them, as Cleve Backster's experiments showed. Animals, too, sense when danger is near. Even humans flinch or draw back from any painful or unpleasant sensation. All living matter seems to have an innate intelligence which manifests in a kind of 'primary perception' of what is good or bad for its particular structure.

Humans have this ability developed to a very high degree, however, we are not aware of it most of the time. Or, because of over-identification with our intellectual faculties, we ignore this sensitivity, and miss what is happening at the other level of awareness.

Have you ever walked into a room and immediately become uncomfortable? Have you noticed how you tighten up, fidget, and feel angry and afraid? Usually, we ignore these feelings and go on with what we are doing. But this reaction is a signal from your nervous system telling you that the atmosphere and the energy in this location is not healthy for you. The same process can happen with people. With one person we feel as though our body is singing. We are loose, relaxed, and comfortable, whereas with another the belly and shoulders tighten, we feel cramped, our breathing becomes unrhythmic and shallow, and after a few minutes of conversation we are exhausted. Yet outwardly, our five senses have not given us any data to support our feeling of discomfort. The other person may be wealthy and educated and have the manners of an English duke, but all the same our feeling persists. Something deeper in us, something more basic and real, is registering negative energy, and communicating this information through the nervous system.

The human nervous system is still very much a mystery. We know that it is the communication system of the body. Through the nervous system the brain gets all its data from the internal organs, and then transmits the appropriate messages back to those organs. It is the means of organismic coordination.

The possibilities of the human nervous system seem infinite. And Count Alfred Korzybski goes so far as to define life as the amount of energy any given nervous system contains. The system seems to function like a cosmic computer, attached to an ultrasensitive cosmic receiving set. A properly trained and sensitive nervous system would not need any external aids to find out any information it desired. A person having such sensitivity would merely have to think of the question or problem and his brain would send out beams to explore infinity and bring back the desired information. He would receive or perceive the answer as a physical sensation.

Unfortunately, most of us are not that developed. We need concrete aids to interpret and amplify the signals that our nerves wish to communicate to us. And that is precisely the function of the pendulum.

It is not the pendulum itself, therefore, which is giving you the answers. It is your inner higher intelligence communicating through the nervous system – which gives you signals. What the pendulum does is amplify the signal and interpret the meaning of the signals through the codes set up between your conscious and subconscious minds.

A good dowser or pendulum operator does not just see his answer in the movement of the rod or pendulum. He also feels or senses the answer – in terms of frequency registrations – in his hand, or arm, or entire body. But this comes after diligent training.

So when a pendulum operator holds his pendulum over an object, or person (for example, medical radiesthesia), he is measuring the interaction of a given force field with his own nervous system.

But, as early experiments with radiesthesia found out, it is not necessary to have the objects or people actually present to get accurate readings. The subject could be hundreds or even thousands of miles away and it would make no difference. Abbé Mermet, a French priest and one of the great pioneers in the field, was able to dowse for water and minerals in Africa, while seated comfortably at his desk in a French village. He simply held his pendulum over maps of the particular territory and wherever water or minerals were present, the pendulum would gyrate clockwise.

Verne Cameron, a famous water dowser, was barred from leaving this country because he was considered a security risk. Why? Using map dowsing technique in a demonstration to admirals of the U.S. Navy, he accurately located the position and depth of all our submarines and submarine bases in the Pacific. Not only that but he was also able to distinguish between American and Russian subs.

How did he do it? There are all kinds of theories and explanations being propounded. One authority, Max Freedom Long, says that it is subconscious phenomena which the pendulum makes conscious. Verne Cameron himself maintains that it is superconscious energies. Others propose all kinds of metaphysical and religious explanations for the phenomena – angels make the pendulum move, or God makes it move, and so forth.

But the importance of any craft or skill is not in the 'why' but in the 'how'. Once the 'how' is known, the 'why' really makes little difference. Theories are only useful to still the insatiable curiosity of the intellect which cannot tolerate uncertainty. The intellect must have reasons.

So to explain teleradiesthesia (detection from a distance) here is as good an explanation as any: The mind operates something like a combination radio or TV receiver and transmitter. A person with a properly trained mind who can concentrate and hold his thought powerfully on a particular object, thought, person, substance, or idea becomes in tune with it. The person touches the thing on its own frequency. Nerve cells begin to vibrate in resonance to it, and this vibration has a frequency (vibratory rate) and wavelength which give it a unique quality, tone, or colour. Isidore Friedman, a master pendulum operator, maintains that these frequencies also have a certain shape. The nervous system then translates this quality and causes the appropriate agreed upon movement in the pendulum.

Now since everyone is familiar with radio band and TV frequencies, this principle should not be difficult to understand. Attention and concentration are the tuning devices of the mind. The transmitting stations are the objects or people themselves, which are continually broadcasting or radiating energy frequencies. The only problems that occur are when there is too much static or interference in either our own heads or in the environment. Then the pendulum readings can be off.

To carry our analogy of radio and TV a step further; we all know that when there are electrical disturbances (for example, storms or lightning in the atmosphere, they interfere with our radio and TV reception. The sound and pictures buzz, whir, and jump around. The frequencies shift and become distorted. A similar thing happens when electrical storms take place within our own minds and feelings, or when certain planetary influences disturb the electrical equilibrium of the mental atmosphere.

To deal with the static in the environment is easy. We just postpone our reading to another time when the situation is more peaceful. Fortunately, these atmospheric disturbances are only temporary. To clear up the static in our minds and feelings, however, is quite another matter. It requires training and discipline. It requires the ability to concentrate and focus. And this is really the hardest part of using the pendulum.

Anybody with a trained mind who has a reasonable ability to concentrate and who is emotionally stable can use the pendulum, and become quite accurate with it in a very short time. Those who do not have these necessary abilities will not do very well with the pendulum. Our advice is to acquire the necessary mental and emotional control first before putting too much trust in the pendulum readings.

If you want, you can use the pendulum to develop these qualities. For as you start to practise becoming emotionally neutral, you will gain more control and poise, and you will notice that your readings become progressively more accurate. In addition, other facets of your life will begin to improve, as well.

In any event, what the science of radiesthesia needs is not more theories but more trained practitioners who can see for themselves what it can do, and who will extend the scope of research into this most important New Age science. Once you begin to operate the pendulum and prove its worth to yourself, you will not be too concerned about the theories behind it. As Edison replied, when asked, 'What is electricity?' 'I don't know what it is, but it's there, let's use it.'

4

Making Your Own Pendulum

Pendulum manufacturing is becoming big business. As the demand increases for accurate and well-balanced 'pendules', as they are called in Germany, more and more concerns are crafting pendulums for profit. Single craftsmen, often radiesthetists themselves, work hours in their laboratory or workshop perfecting yet another type of pendulum; the varieties now number in the thousands, if not tens of thousands. Naturally, the more common, practical, and longer-lasting substances include wood, metal, glass, bakelite, plastic, and amber. Precious metals and other exotic materials not used as often include beechwood, ivory, marble, gold, silver, rock crystal, imported Czechoslovakian lead crystalware, jade, amethyst, lapis lingua, chrysocollo, whale bone, and others.

Most radiesthetists and pendulum craftsmen say metal pendulums should only be used for specific purposes as they respond more readily to certain definite influences. The reason is that metals, as a rule, act as conductors. Thus, the metallic vibrations interfere with accurate readings. However, an iron pendulum is particularly sensitive to magnetic fields while one made of copper is usually susceptible to very minute electric charges. On the whole neutral substances, that is, non-conductors, make the finest pendulums; glass, wood, and plastic are the most popular.

Generally, pendulums are round, cylindrical, spherical, or cigar shaped. Above all, it is vital that they be symmetrical. The spherical shape, it seems, has the advantage of being less affected by the wind. Adversely, the sphere is often insensitive to slight changes so critical in some measurements. Most pendulumists use the cylindrical shape because of its accuracy. Some pendules have a hollow compartment in which to place samples of the substances being analyzed. The disadvantage of this type is the interference from residual vibrations left by the sample. For example, if a gold treasure is being divined with a small nugget of gold inside, and then a chunk of silver substituted, the gold residue may prevent the discovery of the silver.

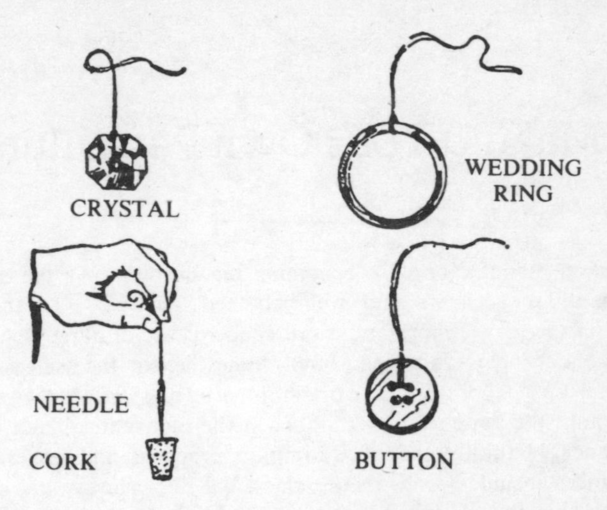

Figure 3. Pendulums made from everyday objects.

The pendulum craftsman must choose a flexible link to the pendulum body so that it is capable of balancing freely. Thread, cord, chain, white hemp, nylon fishing line, black silk, twine, and string are most commonly used. Besides flexibility it must be pliant and strong. Fine nylon fishing line seems to be the cheapest and most durable. Common sewing thread is flexible and pliant but lacks the strength. It usually does not withstand repeated use, breaking within a few months of regular use.

How can you craft your own pendulum right now, at home? Numerous pendulumists have written books in which they describe the making of simple homemade pendulums for beginners. Sybil Leek, the internationally known astrologer and self-professed 'witch' writes: 'To make your own pendulum, obtain a small glass or plastic bead and attach it to a piece of thread, preferably of black silk. An alternative to the bead that is often used is a small ring, such as a wedding or friendship ring.' (See Figure 3.)

The English doctor and radiesthetist, Henry Tomlinson, instructs beginners to 'take a black wooden spool suspended on a black thread, leaving about six inches of thread at the end'. Both Tomlinson and Leek suggest black but not because of any evil connotation such as black magic. The reason is simple and scientific: black vibrates a less disturbing wave than do the rainbow colours, thus cutting down on the possibility of stray interferences.

Dr Howard Brenton MacDonald, a Canadian, writes in his pamphlet *How to Make and Operate A Pendulum*, 'Pendulums can be made from a variety of materials. You can make a simple one out of a piece of simulated pearl, such as a blouse button, tied to the end of an eight-inch length of heavy cotton thread. A pendant shaped earring, also of simulated pearl, makes another serviceable pendulum for beginners. All you have to do is cut off the lobe clasp and tie the ornament to the end of a piece of thread or chain.'

A noted psychic researcher, William J. 'Bill' Finch, describes do it yourself pendulum making in the book *The Pendulum and Possession*. 'You may make your own pendulum by using a needle inserted into a ball of aluminium foil, or a cork, or any substance through which the needle can be pressed. The eye of the needle will provide a place for attaching the string or chain. The point may then become an indicator. We suggest that you snip off the sharp point to prevent an accident ...'

Beverly C. Jaegars, who teaches an accredited course on the pendulum in St Louis, best describes handcrafting your own pendulum. 'A pendulum can be simply a ring or pointed object suspended from a length of string or chain. A locket, pendant, or small fishing weight will serve until you are able to find a more professional type of instrument.'

Two French radiesthetists advocate the use of another type of pendulum. They add a small stick to the weight and string. Henry De France gives an account of how to make this three-part pendulum. 'Take a wooden ball or cylinder weighing from one to two ounces or more – a heavy pendulum is preferable for beginners, then a length of ten-inch pliant and strong thread ending in a loop. Make a hole in the ball and fix the loop in the hole by means of a small wooden peg. Fasten the thread to a little stick about four inches long and one-eighth to one quarter inch in diameter, and wind up the thread on it near one end.'

Another Frenchman, Pierre Beasse, in his treatise, *Dowsing According to the Methods of Physical Radiesthesie* explains his crafting steps. 'The sphere may be constituted by a large marble, such as schoolboys use, made of glass or sandstone and about one inch in diameter which may be found in any department store, or by a wooden ball from a small cup and ball toy. In the ball is a small chink into which is inserted and glued a celluloid slip .08 inches wide, in which has been bored a small hole through which is fastened the suspensory string, the little stick being used to determine easily its length.'

As both Frenchmen state quite clearly this type of pendulum should be held by the little stick between the thumb and the index finger. Then, once a comfortable length of string is found, individual for each operator, the pendulum will start rotating according to the vibration of the radiating object.

There is a variation of this pendulum called a 'hollow pendulum' – similar to the receptacle types mentioned previously – which holds a sample of the substance being dowsed for. It is constructed of a small container of earthenware or wood having a sturdy handle to which the thread or string is attached.

We have crafted our own pendulums which have proved in practice to be extremely accurate and sensitive (See Figure 4). Most timber merchants sell 18-inch long dowels which are ½ inch in diameter. The dowel should be cut, preferably with a jigsaw, into 2¼-inch and 2½-inch pieces. Both ends of each piece should be sanded so that the edges are smooth. Purchase from a hardware store half-inch rubber or plastic furniture leg tips and half-inch rubber-ended tacks that are rounded. Also it is advisable to buy from a sporting goods store 10- to 12-pound test nylon fishing line. Cut approximately an 8- to 10-inch piece of fishing line and tie a knot at each end. Next, punch a hole in the centre of the plastic tip with a tool similar to one used to punch holes in leather belts. Push through one end of the fishing line and squeeze a bit of cement glue on one end of the dowel. Quickly, place the tip over the end of the dowel and let it dry.

Now, you can paint or stain the dowel as you prefer. We stain our dowels with a French walnut stain, but you may use any other stain that pleases you; let it dry, and then varnish it to give it a shiny, well-finished look. After the varnish is thoroughly dry, let it dry overnight, then hammer the round tack into the end of the dowel.

There are, of course, variations of this pendulum. Instead of using a furniture leg tip, use a tiny eye screw, and tie one end of the fishing line to it, using a fisherman's knot. A fisherman's knot is made by slipping the line through the eye, twisting the line around five to ten turns and then putting the end of the line through the small opening of line next to the eye screw. Pull the line tight, and there is your fisherman's knot. Next, you spray the pendulum with black paint and hammer the furniture tack into place.

An easy variation on both pendulums is to paint the colours red, yellow, and blue in on quarter-inch to three-eighths-inch strips from top to bottom. The three colours should be evenly measured, each one-third of the pendulum's length. We will explain in detail the purpose of the colours in a subsequent chapter on 'Improving Your Health' in which we discuss colour healing, colour balancing, and the occult inner bodies.

Whether you choose a bead, a ring, or a dowel pendulum, like ours, the best advice to pendulumists is – craft your own instruments. Usually, it can be done easily and inexpensively and, by the way, has the advantage of providing a pendulum exactly fitted to your personal vibrations.

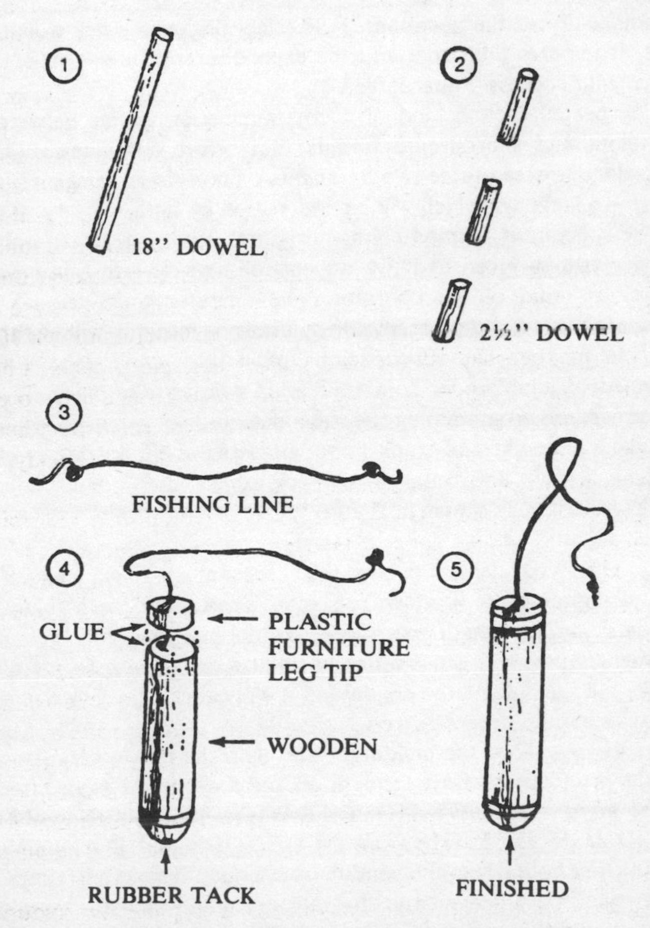

Figure 4. Steps for making your own pendulum.

5

How to Use the Pendulum

Learning to use the pendulum is like learning any other worthwhile skill. It requires time, toil, and the expenditure of intelligence. Given these, your success is guaranteed.

The pendulum is a kind of a communication device between the conscious and subconscious minds. But before communication can take place between these two 'strangers', a mutual language has to be set up – a language which will be understood by both.

The subconscious mind cannot originate anything. It can only act on suggestions given to it by an outside source – in this case the conscious mind of the operator. The operator must tell his own subconscious what the acceptable symbols or motions will be for yes, no, and maybe. The subconscious must fully understand what is expected of it or proper communication cannot take place. We call this procedure programming the subconscious.

Take a large sheet of blank paper and draw the following symbols: vertical arrow, horizontal arrow, clockwise circle, and counter-clockwise circle, as shown in Figure 5.

1. Hold your pendulum over the vertical arrow. Give yourself a 3-inch string length to start with. (Note: There is no such thing as a standard string length with the pendulum. As you get more experience, you will get a sense of what string length to maintain.) While you are holding the pendulum, it will either be still or else it will gyrate randomly over the arrow.
2. Now look at the pendulum and with the power of your mind will the pendulum to move straight up and down in the same direction as the arrow. DO NOT MOVE IT WITH YOUR FINGERS OR HAND. USE ONLY THE POWER OF YOUR MIND AND WILL.

Nine times out of ten the pendulum will obey. This often comes as a surprise to many people who do not understand mental phenomena, but it is perfectly natural. The mind does have the power to affect matter, and this is one of the simplest methods of proving it.

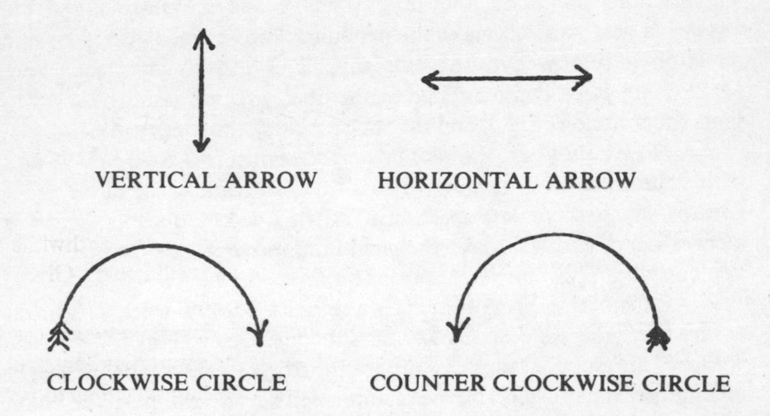

Figure 5. Symbols for setting up pendulum communication.

If you are having difficulty in making the pendulum obey your mind, you are either tired – which is usually the case – or you do not have the necessary mental voltage to make it move. The latter is very rare. Even children have little trouble here. If you have rested and still can not make the pendulum move, then you must begin to build up your mental voltage and power by doing this exercise daily. If you persevere, you will eventually be able to make the pendulum move. The mind can be strengthened by mental calisthenics much the same way the body can be strengthened by physical exercise.

3. Next, hold the pendulum over the horizontal arrow. Again look at the pendulum and with the power of your mind direct it to move in the direction of the arrow – horizontally.
4. This time repeat the procedure holding the pendulum over the clockwise circle, and with the power of your mind, direct it to move in a circular direction that matches the direction of the circle.
5. Do the same with the counterclockwise circle.

Before taking the next step, make sure that you have confidence in your ability to make the pendulum move in any direction you will it to.

6. Once the ability is established in yourself, make the pendulum move in a clockwise direction while holding it over the clockwise

circle. Make it gyrate in a nice smooth circle. As the pendulum is gyrating, address your subconscious. Speak to it out loud. Call your subconscious by name and say, 'When I ask a question and the answer is yes, you will make the pendulum move clockwise, the same direction it is now gyrating.' Or say, 'This movement means yes.' Make your voice commanding. Remember you are issuing orders to your subconscious mind, and the subconscious must obey you.

7. Now hold your pendulum over the counterclockwise circle and with your mind make it move in the same direction as the circle. As it gyrates say to your subconscious, 'When I ask a question and the answer is no, you will make the pendulum move in this direction,' Or you can say, 'This movement means no.'

8. Repeat this programming once a day for a week. This will ensure that your suggestions to your subconscious will take.

Once the subconscious is programmed you are ready to begin making practical use of the pendulum. Before you do, here are some important points to remember.

1. The pendulum will answer any question that requires a yes or no answer. But you must make sure that your questions are phrased in such a manner. It is useless to ask it questions which cannot be answered with yes or no signal. The nervous system has no language with which it can answer through the pendulum.
2. Your mind and emotions must be in a neutral state. You cannot have an opinion or feeling about the outcome of the question if you want an accurate reading. Your feelings and desires will influence the swing of the pendulum. For best results, take the attitude of someone who is reading a gas meter or a voltmeter. You are alert, but not opinionated. You want a true answer even if it goes against your desires. The attainment of this state is perhaps the hardest part of mastering the pendulum, but once attainted you will have an invaluable tool for life. You will have acquired a sixth sense. Your consciousness and awareness will have expanded, and it is an experience that cannot be described in words. Until you have attained a degree of neutrality in your questions, it is wise when checking on areas that are close to your heart, where you have a personal stake in the outcome, to have the reading done by a friend or other person who uses the pendulum, but who has no personal interest in the answer. He or she will be better able to maintain the neutral, objective state so necessary for getting accurate results.

3. Make sure your emotions are calm and stable when using the pendulum. If you are upset, or elated, or in any kind of emotional tizzy, it may put static in your readings. The reason for this is that strong emotions tend to shake the nervous system. The pendulum operates on bioelectric principles. Strong emotions add resistance to the flow of nerve current that is needed to make the pendulum move. Your pendulum will move erratically. It will wobble, jump, and perform interesting St Vitus' dances, but it will distort your readings.
4. Since the pendulum operates on electrical principles, it is very important that you do not short-circuit it while doing a reading. DO NOT HAVE YOUR HANDS OR LEGS CROSSED OR TOUCHING EACH OTHER WHILE DOING A READING. Sit with feet apart and firmly planted on the ground. If you are right-handed, hold the pendulum in your right hand; if left-handed, hold it in your left hand. Keep your other hand at your side.
5. Know what you are asking. That is, have a precise awareness of the meanings of the words you use in your questions. If you do not know precisely what you are asking, the pendulum cannot give an adequate or correct response. Many beginners make the mistake of asking questions such as, 'Will Harriet be happy with Sam', or 'Will I be successful in life.' The problem here is that most of us do not know what we mean when we talk about things like 'success' or 'being happy'. These terms are too abstract. Keep your questions as specific as possible. If by success you mean making a particular amount of money out of a deal, then that is the question you should be asking. This fuzziness that most of us have about words is a serious obstacle to accurate pendulum readings.
6. Once you have programmed your subconscious on the proper gyrations for yes and no, and the suggestions are taken, you can add other refinements to your technique. Using the same procedure just outlined, you can program a clockwise gyration to mean POSITIVE and the counterclockwise gyration to mean NEGATIVE; a clockwise gyration to mean HARMONIOUS, and a counterclockwise gyration to mean INHARMONIOUS. These new terms will have importance in more advanced uses of the pendulum.
7. After you have established your signals and your programming is complete, it never has to be repeated. You just sit down and do your readings.

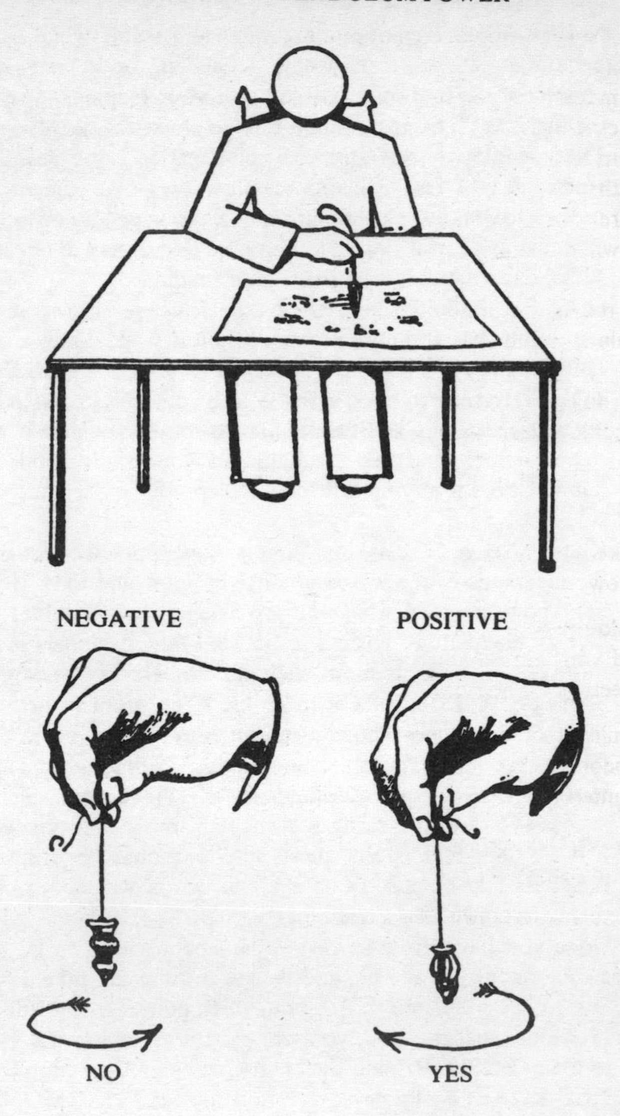

Figure 6. Positions for operating pendulum.

8. As we mentioned, the pendulum operates on electrical principles, and reads radiations. Therefore when you do your readings, make sure you are not around too many electrical appliances – especially of the high-voltage variety, for example, X-ray machines, colour TV's, high-voltage machinery. These machines throw off powerful radiations which tend to throw off the readings, and make the pendulum oscillate erratically, especially when the operator is untrained.

Let's recap the main points of preparation in practical use of the pendulum as shown in Figure 6.

- Sit down with your feet planted firmly on the ground, hands not touching each other; legs not touching each other either.
- Be in a quiet place, with a minimum of electrical equipment in the area.
- Take a blank sheet of paper and draw a vertical arrow, a horizontal arrow, a clockwise circle, and a counterclockwise circle.
- Hold the pendulum over each of the symbols, look at the pendulum and with only the power of your mind, make it move in the same direction as the symbol.
- While the pendulum is gyrating clockwise say to your subconscious, out loud, 'This means yes.' While it is gyrating counterclockwise say to your subconscious, 'This means no.'
- Repeat this exercise every day for a week, to ensure that your suggestions have been accepted by the subconscious.
- You are now ready to ask questions and open up a whole new field of communication with your subconscious.
- Make sure that all your questions can be answered with a yes or no response.

6

Romantic Radiesthesia

The party was in full swing. The pulsating, passioned rhythms of Gladys Knight and the Pips blared from the four large floor amplifiers that anchored each corner of the living room. People were all over the place, dancing, drinking, laughing and buzzing.

Normally, the kind of party I prefer is the small, quiet variety, but I was here at the special request of the hostess who was an old friend. She wanted me to meet some people who were interested in learning more about the pendulum. But till now she was nowhere in sight. I caught brief glimpses of her running in and out of the kitchen with trays of hors d'oeuvres, but that was it. I was pressed up against the wall trying to protect both my stomach and my drink from being emptied by the stray elbows of the dancers.

I finished my drink and was trying to figure out a way to get through the throng to the bar for a refill when I spotted B.B., an old friend. As usual she was surrounded by a group of admiring men. This was not surprising. She was long, lean, and leggy with a finely sculptured face and long black hair, a knockout. When she saw me, she came running over.

'You've got to do me a favour,' she said.

When B.B. asks you to do her a favour in that husky voice of hers, you've got to be superhuman to refuse.

'Sure. What is it?'

'I know you're into all that pendulum baloney and you know I don't believe in it, but right now I've got a problem. I'm about ready to run out and find a gypsy with a crystal ball. In other words, I'll try anything. I've got three of the most gorgeous men I've ever seen in my life ready to take me home. I like all of them. Take out your pendulum, and use your hocus pocus ...'

'It's not hocus pocus B.B., it's perfectly scientific and rational ...'

'Well whatever you want to call it. Use your pendulum to check which man I should go home with tonight.'

'Some problem,' I said. 'Most of the women in this room would give their right eye to have that problem.'

'Don't be a wise guy. Do me the favour.'

'I can't. There are too many people in the room and their vibes will throw off the reading.'

'Leave that to me.' And with that she took me by the hand and led me to the bathroom. Even that was crowded. But that didn't bother B.B. She announced that everyone had to leave because she and I had to do a pendulum reading together.

'So that's what they're calling it these days,' said one of the girls as she left.

Finally, we were alone. I took a few deep breaths and quieted my mind, and then I did the reading. One of the prospects was eliminated completely because the reading showed they were completely incompatible. The other two prospects both showed compatibility, but the positive swing was stronger and steadier over D.G. than over the other. I advised B.B. to go home with D.G. She took my advice, and the last I heard they were engaged.

The point of this story is not to show how you can get into the bathroom with a beautiful brunette, but rather to illustrate the electrical nature of the phenomenon we call love, and how the pendulum can be used to read these electrical radiations. One of the most exciting developments of our times is that love is being taken out of the domain of the poet, artist, musician, and film director and being brought into the realm of science. Man now has the capacity to really understand love.

Since it has now been proved empirically that everything in the universe vibrates, that every object and living entity has a magnetic field, that what we call reality is nothing more than patterns of energy in relationship, that the main characteristics of energy are that it radiates and absorbs, and that life itself could be described as the inter-play between radiating and absorbing energy systems integrated in an energy universe, we have the roots of what Christopher Hills has called the 'Ultimate Science' – radiational physics. Physics is the study of forces. Radiational physics is the study of radiating forces, in other words, all that is. Properly understood, all other sciences become branches or particularizations of this master science.

The problem of love, sex, and compatibility also become branches of radiational physics. We need not approach these areas of life on a 'hit or miss' basis anymore. Once we have the tools for reading and interpreting radiations, an element of predictability comes into our lives. A very welcome addition, indeed.

Who can deny that satisfying and harmonious love and sex relationships are essential to human health and well being? Who can deny that one of the main sources of human misery lies in the forming of wrong relationships? Who has not suffered at one time or another in a bad marital or romantic relationship? Wouldn't it be a boon to all of us if there were some method for reducing the risk of one of these fiascos, or for eliminating them completely?

The pendulum affords us such a method.

What is compatibility anyway? From the standpoint of radiational physics, it is the harmony of two or more given magnetic fields. To get a better idea of what this means, pick up two bar magnets. Put both north-seeking poles together and see what happens. They repel each other. To keep them together requires real effort. It is as if the magnets had feelings and wanted to run away from each other. Their fields conflict. Now take the same two magnets and place the north-seeking pole of one to the south-seeking pole of the other. What happens? The two magnets attract each other. It is difficult to pull them apart. This kind of attraction between two force-fields shows compatibility.

When this kind of attraction exists between two people on the physical, emotional, and mental levels, we say that these people are 'in love'. The converse is also true. When the force-fields between two people are not blending properly, when the polarities are not matched, a state of repulsion or antagonism exists. We say these people are incompatible.

All of us have experienced situations in which we have met a particular person and instantly we felt a kind of irrational repulsion or antagonism. There seems to be no logical reason for it, and very often we do not understand it – after all the other person has not done anything to earn our enmity. But the state exists, and we feel it. We tense up, fidget, scratch, and squirm. Our stomach begins to rumble and perhaps the back of the neck gets hot. What causes this. The aura of magnetic field of the other person is radiating hostile or antagonistic energy. Our bodies are telling us that we are incompatible. As H. Tomlinson says, 'There is no more sensitive detecting instrument than the human body.'

The pendulum is one of the most accurate ways of measuring the interaction or compatibility between force-fields. The advantage of using the pendulum is that you do not have to be actually in the presence of the other person to measure compatibility. You can think of the other person, or tune to them.

Here is the technique. Take a card or sheet of paper and write your name and the name of the person you are checking. Like so:

Your name/Other name

If you are checking two other people, write their names:

Name No. 1/Name No. 2

Get your mind in neutral. Cancel all opinions. Take a few deep, soft, slow breaths and get relaxed and quiet. Hold the pendulum over the two names and ask, 'Are these two people compatible?' When the two people really are compatible, you will get a nice, smooth clockwise (positive) swing. If they are incompatible, you will get a counterclockwise (negative) swing. You can also measure the degree of compatibility or incompatibility by observing the strength and size (diameter) of the circle. Wider and faster swings mean more positive or more negative, depending on the direction of the swing.

Sometimes when you measure two people, you will not get a circular movement. You will get an erratic, dancing, shaking, jumping around movement. If you are really sensitive, you will feel a kind of paralysis or deadening of the nerves in your whole arm. Do not be concerned. There is nothing wrong with you. You are merely tuning into the conflict between two incredibly inharmonious force-fields. When you get this kind of a reading between two people, it usually means that it is disastrous even to have these people in the same room together, let alone enter a romantic relationship. It means total incompatibility – or instant conflict.

When you attain a degree of proficiency with the pendulum, you will have an extra sense. Your awareness of what is happening in the outside world will increase geometrically. If we can call the average person who works with the ordinary five senses an AM radio set, then you will become an AM-FM set. That is, you will be able to receive more stations from the outside. You will perceive things that would usually remain mysterious to the average person. For you will have forged a link with the energy-process universe. The real world.

When this happens to you, there will be a period of readjustment in which you will have to re-evaluate all your old ideas, opinions, dogmas, and antiquated moralities. For all of these were based on an old and incomplete perception of the energy world. Now you will be getting more data, and in light of this data, you will have to formulate new conclusions and new ideas.

You will be saddened to see married people staying together for all kinds of verbal and moralistic reasons when their force-fields show total and destructive incompatibility. You will be appalled at the spectacle of people getting together for all the wrong reasons. People will often marry solely on the basis of a sex attraction, or a money and status attraction without any regard to the compatibility of their magnetic fields.

It is the auric field that will determine whether a relationship will last or dissolve. It is the auric field that will show the happiness and tranquility of a relationship. This auric field should be checked before anyone enters into a serious relationship.

What could be sadder than men and women coming out of a negative romantic relationship, emotionally crippled, physically and nervously exhausted, and very often nursing a permanent grudge against the opposite sex who 'did them wrong'. Most of the time these relationships should never have been born. These couples had no business getting married. If they felt a strong sex attraction, they should have had an affair. In an affair when the sex attraction wears off, so does the relationship, and no one gets hurt.

We are in danger of destroying the planet and ourselves by still applying old moralistic doctrines, conceived thousands of years ago for a primitive and unscientific humanity, to a universe that is now essentially a dynamic, ever-changing, ever-evolving energy process. Our morality must be based on what is if we are to live sanely. In other words, it must be based on modern laws of energy transmission and reception.

The only criterion for sex morality that should exist between two people is the quality and nature of the attraction of their magnetic fields. When two people's fields clash, nature is telling them to stay away. It is useless to invoke dead laws and outmoded creeds to countermand nature's law. All such attempts always lead to sickness and disaster.

Besides being able to check overall compatibility, the pendulum can be used to check sexual compatibility. We all know that two people can enjoy a rich and satisfying sex relationship and still be incompatible when it comes to living or working together. The reverse is also true. Two people can show a high degree of emotional and mental compatibility and have a so-so sexual relationship.

For those people not interested in a permanently structured relationship, the pendulum can be used to check out prospective dates. Blind dates need no longer be blind. If someone has a 'wonderful girl or guy' for you, take the name and check on your compatibility before accepting. It is a proved timesaver.

In checking for compatibility there are certain fine points that are useful for the purposes of precision. Besides checking overall or general compatibility, run a check on each of the following: physical, emotional, mental, and spiritual compatibility. When you get a positive reading on the physical octave of energies, it means the two people will get along physically, e.g., sexually. If there is also emotional compatibility, there is the possibility of a good love affair. If there is mental compatibility in addition to the other two, it is excellent for a living-together or marriage situation. This kind of marriage will usually last as long as the two people keep growing at the same pace. Should the partners grow in different directions, the marriage is in trouble. This is why if you are thinking about marrying someone, you should check on your spiritual compatibility. If this is in harmony, the likelihood of a long-term permanent relationship is present. Both parties will nourish each other and help the other grow.

In relations between the sexes, there is another important factor that is very rarely mentioned in written literature. This is the 'level of Being' of the two people. By level of Being we do not mean spiritual harmony. Spiritual harmony results when two people have the same spiritual ideals, when there is a desire to manifest some ideal that is bigger than oneself. Level of Being is a different matter. A person's level of Being reflects his or her spiritual development or attainment: the stage and phase in a person's spiritual development. It seems that when the levels of the Being of two people are too disparate, a permanent relationship cannot take place. Perfect harmony occurs when there is a similarity in the levels of Being.

It is theoretically possible to measure this with the pendulum, but much more research needs to be done. One of the problems here is that we have no system of measurement with which to gauge such a transcendant quality. But it is possible to compare the levels of two people by asking the pendulum if the levels are in the same octave. Isidore Friedman is doing much original research in this area, and when he publishes his findings, the results should be enlightening.

7

True Stories of Pendulum Power

Through our research and personal experience, we have gathered together a few stories which tell of the pendulum's uses in remarkable or unusual situations. Some of the stories may sound like 'big fish' stories, too tall to be true. In fact, they are all documented and can be checked out. Pendulum Power has many possibilities, some of which are extraordinary.

Fool's Gold

In the early part of the twentieth century in the lovely Swiss village of Sedeilles, Abbé Mermet revealed yet another of his pendulum powers. His original purpose in the quaint community was to dowse for a drinking water supply. But, after hours of searching, he declared to the head of the town council that it would be far easier to find gold than water.

The council head was somewhat shocked at the good abbé's assertion and demanded, with due respect, that he show the gold to him. Not far from where the two stood some harvesters were working in the hot, late summer sun. Mermet suggested the gold was on one of the harvesters. They went to check the pendulum, it indicated gold on one man's person. The hearty fellow rebutted mockingly, 'I don't know where this gold could be with the clothes I've got on.'

A number of people gathered around and soon they all demanded to see whether or not gold was indeed on the harvester. He, thus, stripped to the waist. Wearing only a pair of trousers Mermet still insisted there was gold on him, and the man continued to deny it. Finally, the man wanting to get back to work and be done with the monk's foolishness demanded, 'Well, then, if you know where I have this gold, take it.' Mermet immediately placed his hand on the man's belt buckle saying, 'What about this, isn't that gold?'

Suddenly, the harvester's memory flashed back to an event which, until then, had been wiped completely from his mind. When he was called to war in August of 1914 his mother had sewed a gold coin in the belt of his trousers in case he needed emergency money, but as he was fortunate to always have just enough, the need never arose. Upon his return at the end of the war, he forgot the gold coin altogether.

A Missing Child

In the spring of 1934 a truly extraordinary article was published in the *Geneva Tribune* under the heading 'Disappearance of a Child Explained by Teleradiesthesia'. It seems that in the late autumn of 1933 in Valais, Switzerland a six-year-old boy disappeared without the slightest clue to his whereabouts. Nearly every member of the tiny village joined in the thorough search of the town proper as well as the surrounding countryside. Not even a trace of the boy was found.

The town's mayor wrote to Abbé Mermet asking for his assistance. Using his pendulum he studied the case, and then wrote the mayor an incredible explanation of the boy's strange disappearance. He confidently announced that the small boy had been carried off by an eagle into the mountains. Also he made specific reference to the bird's great wing span and two locations where the giant bird dropped the child while the eagle regained its strength.

With a group of friends, the boy's father, M. Baloz checked the first spot. Nothing was found. They were prevented from checking the second spot because of a heavy mountain snowfall. Everyone concluded prematurely that the abbé was mistaken.

However, two weeks later, the snow melted. At the precise spot indicated by Mermet, a band of woodcutters came across the body of the small boy, badly mangled and torn. They concluded that the ravenous eagle had stopped devouring the boy when the heavy snows came.

Also eyewitnesses observed the boy's clothes and shoes to be clean. Obviously, the powerful eagle must have carried the boy into the high mountains. The boy's father wrote to Mermet on 18 March 1934. 'Now that the body of my poor boy has been recovered, it is our duty to thank you for so kindly helping us and giving us such precise information. Everything has been confirmed. It is now certain, as you said in your first letter, that the poor boy was carried away by an eagle which did not stop in its flight until it reached the mountain heights at the two places you had indicated, and where the body was eventually found. We also observed that the boy's clothes were as clean as they were on the morning of his disappearance. You were the only person who really knew and understood what had taken place.

Please forgive us if we appeared to be very doubtful about your indications. Several eyewitnesses in Sierre declared having seen, on the same day, an enormous eagle flying in a direction towards the north. Again thanking you, etc., L. Baloz.'

Dowsing in Combat

During the Vietnam conflict, Robert McNamara, the Secretary of Defence, asked for ideas which would help solve difficult military problems, such as locating underground tunnels, landmines, and ammunition dumps.

Soon after the US Marine Corp began its investigation into dowsing. Major Manley, along with several of his fellow officers put the radiesthesia techniques to work. They chose the Southeast Asian village at the Quantico Marine Base and went over the area thoroughly with their dowsing devices. Their accuracy was staggering in detecting tunnels, false wells, buried communication wires, and pits.

Louis J. Matacia, an accomplished dowser and a private citizen, instigated the original marine investigation and kept it alive by writing numerous letters to top brass, including General Westmoreland himself. Apparently, Westmoreland responded, for gradually Matacia heard reports from the general's staff that marines were locating all kinds of places and things including tunnels, caves, caches of food, ammunition, and secret messages buried in bamboo tubes. Military leaders were impressed by the concrete successes.

And then in October of 1967 combat dowsing got the rave review it needed. Hanson Baldwin, a *New York Times* military reporter wrote an article, 'Dowsers Detect Enemy Tunnels'. Hundreds of newspapers, big and small, republished the article in condensed form. Magazine articles, television newscasts, and radio reports flashed from coast to coast. Dowsing became a handy tool of defence against the enemy.

Black Gold

The year was 1943. World War II was in full swing. The need for oil was great. A well-known wildcatter, named Ace Gutowski, who was backed by the Fox Brewing Company of Chicago started searching for oil in central Oklahoma.

Ace contacted a farmer, who lived in the area of West Edmond Field, central Oklahoma, whose name was J.W. Young. His dowsing device, or doodlebug as he called it, was a most unusual type of pendulum. Young used a small bottle covered with goatskin and filled with mysterious substances, the composition of which he refused to reveal. The bottle was suspended from a watch chain. He held it out from his body over a spot on the ground. It swung east-west over salt water and north-south over oil.

A geophysicist watched a demonstration of Young's doodlebug in an Edmond restaurant, in 1944. The bottle pendulum was held over a specimen of sand. The scientist claimed the device worked well but was amazed at the results since Young suffered from palsy and there was a noticeable tremor in his outstretched hand.

Despite the handicap, Young showed Gutowski where to drill. They discovered the largest oil pool found in Oklahoma in twenty years. It was even more remarkable that the pool was under a geological structure not detectable by conventional geological or geophysical methods.

Dowsing for Submarines

In early 1959 Verne Cameron, a professional Californian dowser, contacted the United States Navy and told them that using a map and a pendulum, he would be able to locate the entire submarine fleet. On 18 March 1959 Vice Admiral Maurice E. Curtis wrote a letter to Mr Cameron:

> I am advised you believe you may be able to tell the location of all submarines in the world's waters – and their nationalities – by a technique which is called 'Map Dowsing'. It has been suggested you be given an opportunity to confirm your ability on the subject of submarine detection and location at a naval establishment close to you.
>
> Please be assured I should welcome a demonstration by you at a place of your choice on the West Coast.
>
> If you will communicate with me about your itinerary for the next month or so and your choice of a place where you can demonstrate this ability. I shall be pleased to arrange for a test.

Soon after receiving this letter Cameron met with the Vice Admiral in Southern California and demonstrated his map dowsing technique for the Navy brass. Within a few minutes Cameron located every single submarine to the absolute amazement of those present. Then, surpassing his feat, he located every Russian sub around the world. Despite the success of the test, Cameron did not hear from the Navy or United States Government until years later under different circumstances.

Cameron had been invited by the South African government to come to their country to discover various natural resources using his pendulum. When he applied for a passport, to his great surprise he was turned down. He investigated the reason for the denial and found out that the Navy had contacted the C.I.A. about his submarine dowsing ability. The intelligence agency labelled him as a security risk and did not allow him the freedom to travel outside the country for fear he might reveal top secret military information.

8

The Pendulum, Your Work and Your Career

Choosing a Career

When Bob D. got involved with the pendulum, he had already been a successful dentist for more than ten years. He was unhappy, however, and what was even worse, he couldn't figure out why. Everyone told him he was crazy for feeling miserable; after all he had what everyone longs to have – money, position, respect, a nice home, a nice wife, two cars in the garage, and so forth.

Bob began to think he was crazy, too. He went to a psychiatrist, had a few extramarital affairs, smoked pot, and began to drink heavily. Nothing helped.

Until now Bob had been using the pendulum to locate bad teeth in his patients (see chapter on medical radiesthesia for the technique) when it finally occurred to him that perhaps he could use it to locate the source of his problem. By using a process of elimination, he found that he was in the wrong career. When he asked the question, 'Should I be practising dentistry?' the pendulum gyrated wildly negative. He repeated the experiment every day for a week at different times of the day, in different environments and in different moods. The reading was always basically the same. Negative.

Then he began making a list of various professions and activities and checking which of these was best suited for him. Again by a process of elimination he discovered – to his great surprise – that singing was his calling. As he mulled over this strange reading he remembered how he first decided to practise dentistry. His parents had really decided for him. And then he remembered that as a child he had fantasies about singing on stage which he had rigidly repressed. He found it hard to believe that these fantasies and dreams were some kind of inner voice telling him what to do.

In the light of his pendulum readings, Bob began to take singing lessons. (He says that his singing teacher was also selected by means of the pendulum.) Lately, he has been appearing at amateur showcases around town, and eventually he hopes to break into the night club circuit full time. He still practises dentistry while building his singing career, but he is much happier.

It seems that everything in nature, man included, has its function and purpose in relationship to the whole universe. As minute parts or cells in this vast multiplicity, we as human beings have to find our place and our particular function. This function is determined by both our inner and outer structures, by the psychological and physiological qualities and aptitudes that we are born with and those that we acquire by training. All of these aptitudes are reflected in our magnetic fields. Under normal circumstances, it should be easy to find our calling. We would simply naturally gravitate towards it. We would attract our calling or profession the way a magnet attracts iron fillings. And this was true in ancient times when people lived closer to nature and were more in touch with their feelings and their intuitive faculties. For example, a person would feel the need to write or to paint; one child might just naturally be attracted to the tribal medicine man and hang around him; another child might naturally get his pleasure by following the warriors around, and so on. In this way, the youngsters would be led to the appropriate profession.

Today it is quite different. There is an anti-natural order of cultural overlay that clogs and pollutes our natural instincts and propensities. It is considered a tragedy if a middle class youth chooses to be an automobile mechanic, even though he may love the work and have a real gift for it. We worship 'success' and believe that it lies in making a lot of money in any way that we can. As a result millions of children are blindly sacrificed to this demon Moloch called 'success'.

The development of the intellectual faculty in man has made possible all the wonderful scientific and technological comforts that we now enjoy, but it has also cut us off from reality, that is, the dynamic energy universe. We choose our careers, lovers, and lifestyles by a curious and spurious system of words, ideas, and verbal concepts based on false-to-fact assumptions of the nature of the universe.

We live and die by our logic, and yet we do not realize that logic is only as good as the assumptions on which it is based. The most gifted, brilliant logician will be completely wrong if he starts his train of thought with a false axiom.

Nature is 'alogical', that is, beyond logic. To master nature, to merge with it, to harmonize with it requires measurement rather than logic and ideas. We must build our logics on empirical measurements of the energies, forces, forms, and functions of a non-static universe.

How should we choose our life work? Should we choose a profession simply because it will earn us a lot of money? Or because someone in our family feels that this is what we ought to be doing? Or just drift into the easiest thing that crosses our path?

Or maybe we should listen to that deeply buried, but sacred inner urge, that deeply repressed heart's desire which is God in us seeking expression? Those of us who are in touch with this part of ourselves do not need the pendulum to choose our life work. We already know what it is. But for those of us in whom the moil, turmoil, and confusion of the outer has drowned out the inner urge (and this includes the great majority of us), the pendulum is an excellent method for choosing our careers provided we learn how to use it properly.

To select a career with the aid of the pendulum, first write down all the things you have ever considered doing. Include on this list your childhood fantasies, suggestions from other people, things that you know you have an aptitude for, and those careers which appeal to you because of the money or prestige, or both.

Get very calm and neutral. Breathe quietly, slowly, and deeply. Relax. Empty your mind of all extraneous thoughts, and of all opinions and feelings. Make sure you are not tired when you run this check. (It is preferable, therefore, to do it in the morning or at a time when you are not pressured.)

Then go down each item on your list and ask the pendulum, 'Is this a wise choice as a career for me?' Many items will be eliminated immediately. At the end of your first session the pendulum may have said yes to several possibilities. Recheck these again at another time. The pendulum will then eliminate a few more. Recheck again another day. Eventually, only one or two careers will be left. These are the areas you should be concentrating on. These are the areas in which you will most likely achieve success, for these are the things for which you have the most aptitude.

What a person does is a function of what he or she is. The person's suitability for a particular line of work depends on the structure of his or her auric magnetic field, and this is exactly what the pendulum is measuring.

Selecting the Right Job

Getting a job is a tricky business. It is not that getting a job *per se* is tough. Even in bad times most people, if they persist, can find some kind of work. But the tricky part is finding a job situation that is right for you. Here, other factors come into play.

We want, first of all, work that suits our nature and aptitudes. We want to work in congenial surroundings and a harmonious atmosphere. We want to be adequately compensated for our labour, and many of us want an opportunity to grow in a job. Of course, everyone is different and unique, and each of us may find that different factors are more important than others.

Some people want jobs that afford them a lot of freedom; others are only interested in how much a job pays; others seek the right atmosphere. This is all well and good, but the problem is how do we find the right kind of job situation that meets our personal needs and long-term goals?

Like so many other important areas of life, we approach the job market on a hit or miss basis. This is why everywhere you go, you see people who hate their work, their bosses, and their co-workers. Enough of this kind of bad feeling, even if it is not evident on the surface, can wreck or seriously damage a company's productivity and sales.

Here is how you can use the pendulum to choose the right job for you. If you are in a position in which you have a choice of various concrete offers, simply write down the names of the companies on a sheet of paper, and hold the pendulum over each name. Ask, 'Is this place right for me at this time?' Ask this question over each name. If you get positives over each of them or negatives over each of them, ask, 'Is this job better than the others for me?' One name will produce the smoothest gyration. That one is the job you select. If you continue to get negative readings on all the names, it probably means that all of these jobs contain factors that are not good for you. Look for other possibilities.

After you become proficient with the pendulum (this takes about a year for the average operator), you will understand why the pendulum does not operate by logic. A job situation that seems ideal from a logical viewpoint, that is, money and prestige, may contain hidden factors that completely neutralize these positives. The work may be back-breaking and gruelling; the boss may be a tyrant; the company may be headed for bankruptcy in the near future; or may be in the process of being taken over by a big conglomerate who will close down your department, to cite just a few possibilities.

I have seen things like this happen dozens of times. What we see of a situation is only a small part of what is actually happening. Our physical senses record only the tip of the iceberg of the dynamic-energy process that we call reality. What the pendulum does is increase and widen our range of perception into the iceberg, thus giving us more data on which to base a decision.

Executive Decision Making

Every human being must take a certain number of decisions each day. But the kinds of decisions busy executives make usually involve that precious symbol of human energy, skill, and cooperation – money. One wrong decision by a high-level executive can ruin a company; one good one can earn millions. So it is not surprising that executive decision making very often becomes a fearful ordeal.

With the use of the pendulum, however, this fear is eliminated. The executive goes about his or her decision-making process in the usual fashion, collecting all the facts, weighing the pros and cons of each possible course of action, and then if he had any doubt about which would be the best choice, the executive would just ask the pendulum. It makes no difference how complicated the decision or the problem to be solved is. The only consideration is to be sure to phrase the question in such a way that it can be answered by a yes or no response (see chapter on 'How To Use the Pendulum').

Hiring Employees

A friend of ours runs a chain of petrol stations. For about a year he was having a serious problem. His employees were stealing from the till. At each of his stations there was always a slight shortage at the end of the day. The shortages were small enough so that they could be explained away by the employees, and besides it was difficult to pinpoint which person was doing what.

All kinds of solutions were proposed to him, none of which fitted in with his means or his operation. He could computerize operations, but that would be very expensive; and moreover it would not stop the cause of the stealing, it would just make it easier to pinpoint the source. He could hire a detective to do undercover work, but that too would be expensive.

He discussed his problem with Isidore Friedman, a master pendulum operator. Isidore explained to him that his problem was that he was hiring the wrong kind of people. To eliminate the stealing he had to make sure that the people he took on were inherently honest and responsible. 'How could this be done?' the businessman wanted to know. Isidore showed him the pendulum and did a few demonstrations. He checked each of the employees and told our friend which ones were most likely to be involved in theft. Sure enough those employees who were selected were working at the petrol stations that consistently had the most shortages.

The businessman hired Isidore as a consultant, got rid of the problem employees, and checked out all new job applicants with the pendulum. Within a few months his theft problem was solved.

Ralph Waldo Emerson used to say: 'What you are thunders so, I cannot hear what you say.' By this he meant that in the energy-process universe, there are no secrets. Every thought, every mood and feeling, every quality and ability, every dishonest or honest motive has a vibration of a specific frequency. To those who understand vibrations, like Emerson and other seers, these qualities are as plain as day.

Criminal as well as good propensities vibrate in a person's aura. No amount of speech, college degrees, fashionable clothes, or charm can disguise it. When we confront one another in the real world of energy, we are all naked.

When more people learn to register energies, whether through using the pendulum or other means, we will have an effective method of controlling not only crimes of theft but many other crimes as well.

The pendulum is also useful in selecting the better of two job applicants. Every personnel manager has to make some tricky decisions in his work. Very often he gets two applicants with equal qualifications and equal experience and past credits. What to do? With the pendulum the decision can be made in seconds.

Using the pendulum in personnel management can save industry millions of dollars a year. This is especially true with jobs that require an expensive training programme. Many sales jobs fall into this category. The company must invest time and money in training the employee, and it must make sure that the employee has aptitude and the stability to stay with the company when the training period is over. All the personnel manager has to do is hold the pendulum over the job applicant's name and ask, 'Is it wise for us to train this person?' or 'Is it wise for us to invest in this person?'

There are many other uses of the pendulum in this field, and the personnel manager who acquires experience with the pendulum will no doubt discover many other applications and techniques.

9

The Pendulum: Door to the Infinite

All through man's recorded and unrecorded history, whether he shivered in the mountain caves of Tibet, or roasted in the hot desert sands of the Middle East, or ate the heart of his enemy in the frozen tundras of the northlands, an urge deeper and more powerful than even the urges for food, shelter, and sex, gnawed at his insides. How could he manifest his inner visions of what could be and should be into the outer reality of raw elemental nature? How could he bring that which was within him to the without? In short, how could he grow into something bigger and better than what he was?

This urge to grow, to BE more, was expressed in many ways in man's historical struggle with it. All kinds of methods were tried. Some men expressed this urge in conquering other countries and provinces, either personally or by sending proxies to follow strong leaders who did the conquering. Some expressed it by amassing great wealth, or great fame, or great pleasures. For some these accomplishments brought temporary fulfillment, but always afterwards there still remained that irrepressible, deeply buried itch to grow – to be and become.

Every need, every legitimate desire creates a magnetic vortex that attracts the fulfillment of that need. And so it was with humanity. Through the ages the more gifted and advanced members of the race, through what Korzybski* calls the 'time-binding faculty' of man, evolved a science of growth. Knowledge, systems, methodologies, and practices were synthesized into a science of manifesting inner potential, which were known only by a few, as it is with most sciences.

* Alfred Korzybski, the late Polish mathematician, engineer, and philosopher, was the founder of the science of general semantics. He also formulated the famous 'theory of time-binding' in which he proves that man differs from animals in that one generation can start where another leaves off. See his *Manhood of Humanity* for a more detailed explanation of his theory. It is a classic.

They were called by various names in different cultures and times, and were cloaked in various sorts of mysterious rituals, the import of which only the initiated understood.

Here we have the origin of the 'Ancient Mystery Teachings' or what has been called the 'Ageless Wisdom'. And it was these teachings that formed the inner or esoteric side of all of man's religions.

According to Isidore Friedman, the inner meaning of all religions, their true import to humanity is that they presented a technology for living. They were originally intended to be a science of life presented in a way that the people of certain time and clime could understand and relate to.

As the religions grew older, as times, customs, and conditions changed, the outer garb of their rituals no longer had meaning or relevance for the people, and a new 'life technologist' or 'spiritual scientist', or avatar would arise and develop new and more timely teachings. None of them really started anything new, of course. All they did was to reclothe the old wisdom in more fashionable dress.

Today humanity is entering a scientific age. The outer garments of the old religions have lost their relevance to the majority of thinking people. But the quest for the Ageless Wisdom, for the Science of Life and Growth which underlie our religions is as strong as ever. This quest goes under many labels: parapsychology, occultism, metaphysics, mysticism, borderland science, mental culture, radiational physics, biomagnetronics, and magic, to name a few.

Some of the fundamental axioms of all parapsychological research are as follows:

1. There is a science to growing and changing one's being.
2. Man has powers and potentialities that are latent and dormant.
3. Man is destined to grow into something infinitely greater than what he is now. As Nietzche has said, 'Man is a bridge across the abyss, a dangerous crossing from the animal to the divine.'
4. Man has a definite place and relationship in the cosmic scheme.
5. This relationship can only be found after he has developed the necessary faculties and aptitudes to probe into the structure, function, and relationships of the cosmic process.

Parapsychology, therefore, embraces all the old attempts of developing man's faculties. It seeks to apply scientific methods to such diverse fields as astrology, numerology, Qabalah, yoga, palmistry, healing, ESP, magic, telepathy, reincarnation, clairvoyance, and others. Since the pendulum is an extrasensory device, it is a valid area of research in the field of parapsychology. More important, however, it becomes a tool through which we can explore all the other borderland sciences.

Choosing the Right Books

People usually get their first exposure to parapsychology and the occult sciences through books. Either they pick up a book that intrigues them, or they meet a person who intrigues them, and who, in turn, recommends certain books. There are many excellent books on parapsychology and the occult; however, there are also many confused and incoherent ones. Holding your pendulum over any book that you are thinking of buying is a good method for discriminating between buying one book rather than another.

In choosing a book, there are other factors to consider. You can have two basically excellent books, and one will register more positive than another with the pendulum. This is because of the rapport and the resonance you may have with the particular author, which is very important. For when you read you are actually tuning in to the author's mind, and when that mind is more or less compatible and harmonious with yours, you will get much more out of the book.

Also two authors of equal capability and knowledge will approach the same subject from different angles. They will give a different presentation of the subject. Very often you will be resonant to one type of presentation and not another at a given time. This, too, will be picked up by your pendulum.

Choosing a Teacher

When one is exploring unknown territory, or when one is taking a trip to a far country, it is wise to have an experienced guide who is very familiar with the territory to be traversed. The same is true when one is exploring the unknown territory of one's inner world.

Finding the right teacher or guide for you is of paramount importance to your inner growth. Occultism is meaningless – may actually be dangerous – if it does not lead to inner growth. By the very nature of the field which deals with the unknown, or with forces which are not readily perceptible by the physical senses, there are no objective guidelines for determining the quality and the ability of a self-proclaimed teacher. There are no licensing procedures in this field, nor can there be. Who, after all, would be responsible for setting the standards?

In choosing between a qualified and unqualified teacher, the student is often left only to his or her intuitive faculties. When these are sharp and well developed, it is all well and good. When they are not, disastrous consequences often ensue.

How does one find a good group or teacher out of the seemingly multitudinous possibilities? Use your pendulum. Take ten possibilities and check them with your pendulum. Over each name ask. 'Is this the right teacher or group for me?' If you get more than one positive reading, ask, 'Is this one best for me?'

You will often get surprising results. You will find that some teachers of great reputation will register absolutely negative for you, whereas another one who is relatively unknown will be positive. This does not mean that the other is incompetent. It just means that one teacher has more to give you at a particular time. At another time, and at a different phase in your development the other teacher may be right for you.

Astrology

It is an axiom among physicians that correct diagnoses are essential to rational and effective treatment. The same holds true with the psychological ailments that all of us have, at one time or another. Doing an astrological reading is like X-raying the psyche. It reveals the structure of the psyche in clear, concise, symbolic form. The astrological chart, which can be compared to a developed X-ray negative, is one of the most effective diagnostic tools known to man at this time. And herein lies its main use. As a predictive instrument, however, it leaves much to be desired, and in fact seems to be mostly invalid. The reason is that there are so many variables in human behaviour – so many factors that cannot be computed – that any serious attempt to get predictive accuracy degenerates into a guessing game; or worse yet, it masks hunches in astrological symbolic clothing.

Psychological health implies a knowledge of oneself. Astrology is one method for gaining this kind of knowledge. Once this is attained, the person can begin to work within his own structure, discovering the limits of his function – his strengths as well as his weaknesses – and he can begin to consciously work on balancing them. Once a person knows what is going on inside himself and why, he can begin to eliminate undesirable traits and strengthen others that he wants to develop.

It is not our intention to present a treatise on astrology here. This would require volumes in its own right. But essentially, there are three factors in a horoscope: the planets which are energy sources of a specific type; signs of the zodiac which modify and modulate the energy of the planets; and houses which indicate the field of experience in which these energy-functions will manifest.

An astrological chart is basically a mathematical statement – an equation – showing the relationships of the various factors in a person's psyche as they are revealed by the relationships of the symbols.

It is well known among astrologers that a chart is only as good as the astrologer who interprets it. Thus five astrologers can draw up the chart on one person and come up with five different interpretations. The same is true of doctors, psychiatrists, and chiropractors. When a patient comes to them with a health problem, there often will be variation in the diagnosis. This does not mean that they are wrong. It means that in making a diagnosis each doctor interprets the data from the laboratory tests and physical examination from the standpoint of his own temperament, background, and psychological outlook.

The point to remember when choosing an astrologer, doctor, lawyer, priest, or other consultant is whether this person's particular and unique approach is valid and right for you. The important qualities to consider in this area are professional competence and vibrational harmony. The most competent professional may be vibrationally out of tune with you, thus making him less able to help you. You will not be as receptive to his advice. He will not be as insightful and sympathetic to your problem.

There is another problem that has been plaguing astrologers for thousands of years. Most charts depend for their accuracy and validity on correct and exact birth data. The exact minute of birth gives the astrologer the exact degree of the 'rising sign' or 'ascendant'. A difference of even four minutes can change the shape and structure of the horoscope, according to some systems of astrology. Now the problem is this. Most people do not know their exact moment of birth. Many think they do because it says so on their birth certificates, but these are known to be notoriously inaccurate. The hospital staff members responsible for recording the time of birth are often overworked and harried; they are usually not familiar with astrology, and it is usually the farthest thing from their minds. Very often they put down the approximate time of birth, which is acceptable for administrative purposes but totally useless for astrological ones. Unless the mother of the child is conscious enough at the time of birth to record the time and remember it (and this is rare) or unless there is a family member present who can record the exact time, it would be rare for anyone to know his or her exact time of birth.

The problem has been circumvented in a variety of ways. Some astrologers use the technique called 'rectification'. They trace certain well-known events in the client's life to see where they must fit in the horoscope pattern; from this they deduce the degree of the rising sign. Many astrologers find this unsatisfactory. There are too many variables in such a method. People's memories are not always accurate. Some events can be caused by a variety of factors, and are not necessarily the factors used in rectification.

Other astrologers dispense with the houses altogether. They feel that if a thing is dubious it is better not to use it. Other astrologers draw up solar charts or flat charts, or they use Ebertin's system of Uranian astrology.

Then there are those who simply close their eyes and cast the chart based on the time of birth given to them from the birth certificate.

With radiesthesia we have a method for determining the exact degree of the rising sign without knowing the exact moment of birth. We simply hold the pendulum over the horoscope and ask, 'Is this person an Aries ascendant?' 'Is he a Taurus ascendant?' and so on until the pendulum reads positive when the name of the person's correct ascending sign is mentioned. Then it is only a matter of checking the exact degree rising. Use the same procedure as you did for obtaining the exact birth time.

To find the exact degree rising, make a little chart. At the top of the chart, write the person's name. Then write the numbers one to thirty. Hold the pendulum over each of the numbers – think of each number as you do so – and ask, 'Is this the degree rising?' The pendulum will swing positive when you come to the right degree.

This could prove to be a fascinating area of research for astrologers. Note that you should attain a degree of proficiency with the pendulum before using the data in chart interpretations. Test your findings privately with the behaviour and character of the native.

Another problem with astrology is that it is only a diagnostic tool and not a system of therapy. In simpler words, astrology as it is being practised today, can only show where and how a problem is operative in the psyche, but it cannot, (nor does it pretend to) show the means and the methods to accomplish the correction of the problem. Therapeutic astrology is still a science of the future. The use of the pendulum, however, brings us one step closer to a science of therapeutic astrology.

When you analyze the planets closely, you find that each of them represents a root energy or principle that exists within nature and man. Health consists of the harmonious and rhythmic cooperation of all these energies within us. The moment there is an imbalance, the moment there is discord or disharmony between any of these energies, disease results. The disease may manifest physically, as when different organs refuse to cooperate with each other or get unbalanced; or it may manifest emotionally-mentally as when different psychic energies are in conflict or unbalanced. Because mental disease is usually the cause of physical disease, mental-emotional disease is far more damaging both to the sick person and to the environment.

By using the pendulum we can check the planetary energies in which we are deficient, and we can consciously add these energies to ourselves. It is similar to checking for emotional-mental vitamin deficiencies. The effect of an energy deficiency in the psyche is just as important as the effect of a vitamin deficiency on the physical body.

In astrological symbolism we have ten psychological energies within us. We call them planets. These are as follows:

Sun The principle of will, purpose, and the urge to be.

Moon The subconscious mind, memory, the principle of habit formation.

Mercury Logical thinking, the ability to communicate mentally.

Venus The principle of attraction, the urge to relate with others socially and through the affections.

Mars Physical energy, courage, the urge to DO on the physical level.

Jupiter The urge to expand on every level.

Saturn The principle of discipline, order, time, measurement, efficiency and limits.

Neptune The urge to envision a more perfect state of affairs; spiritual ideals, impersonal love, and devotion.

Uranus Originality, the urge to the new and the modern: The need to break through old traditions and manifest one's uniqueness.

Pluto The urge to eliminate and transform that which is harmful to the organism.

The astrologer or astrology student can make a list of the planets and hold the pendulum over each one in turn asking, 'Am I (or is my client) deficient in this energy?' The pendulum will swing positive over the planet in which deficiency in energy exists. Let us say that the pendulum indicates that the person is deficient in Venus energy. It means that the person needs to enter into a love relationship or relate more socially, or get into some artistic form of self-expression to balance his psychic energies and restore emotional and mental equilibrium.

Let us say that the pendulum indicates a Saturn deficiency. This means that the person is suffering from a lack of discipline, order, and efficiency. He must consciously work at bringing more of this into his life if he is to restore emotional and mental equilibrium.

The beautiful part of this method is that we can keep a constant watch on our progress. Also when we use this method we find that whereas at one time Saturn energy was needed, another time may show a Venus deficiency or a Mars deficiency. Our psychic needs, just like our dietary needs, are constantly changing and shifting. The pendulum gives us a tool for monitoring them.

The Pendulum and Meditation

We define meditation as a prolonged, yet relaxed, concentration of the mind on one idea, one object, one problem or one goal. It is at the same time a mind-training device and tool of a trained mind. It is the instrument unsurpassed for creative thinking, problem solving, eliminating psychic blocks and obstacles, and the ultimate manifestation of a person's inner potential into outer reality.

As we can readily see, meditation is much more than just sitting around chanting mantras and staring at one's navel. It is unfortunate that so many people confuse certain specific meditation exercises and practices with the art of meditation and mistakenly assume that mere exercises are all there is to the art of meditating. This is akin to assuming that one song or the ability to play one instrument is all there is to the field of music, or that detective novels are the only kind of fiction that exists.

There is no mystique or mystery to meditation. Anyone who has watched an artist or writer mull and chew over a creative problem, or participated in an executive discussion, or thought deeply and persistently about any kind of personal difficulty has unconsciously been meditating. Meditation as a tool of the mind, however, is much more effective when you do it consciously and when you can turn it on and off at will.

Meditation techniques vary with the purpose and the function that one wants to achieve. Basically, all forms require breath control, a relaxed body, and a quiet mind. The techniques to achieve this state vary. There are meditations on the breath, on the body, on the feelings, on the feet, on thoughts, on ideals and transcendant qualities such as beauty, love, God, and so on.

Since each of us is a unique entity, with different structures and temperaments, it is not surprising that each of us will do better with different meditations. The pendulum can be used to determine which one is best for us. Aside from selecting the particular form of meditation you can use the pendulum to determine the length of each meditation session and to locate the best place to meditate.

Mental exercise is similar to physical exercise. You have to work within the limitations of your structure or you will strain your mental muscles. This is why meditation sessions should not be indulged in too long, especially when you are first starting.

Make a chart of a clock with sixty minutes around. Hold your pendulum over the five and ask, 'Can I meditate for five minutes today?' If it swings positive, ask: 'Can I meditate for ten minutes today?' If it reads positive, keep increasing the amount of time in your question until the pendulum reads negative. This will show the limits of safety for you.

When you begin to train on the higher mental levels, you will find that atmospheres become very important. Every place has a mental-emotional atmosphere, which is determined by the frequency rate or vibrations of the people who habitually visit or live there. If too many people with low-quality thoughts or moods are in a place for any length of time, their atmosphere or vibrational emanations tend to linger, and the place is not conducive for meditation. Certain areas, such as hospitals, mental institutions, and tube trains are loaded with negative vibrations and negative astral entities and are unfit as areas of meditation, for the average beginner. If an untrained beginner were to meditate in such a place, there would be a good chance that these entities would enter the meditator's aura (magnetic field – psychic skin) and possibly cause sickness, loss of energy, or obsession. It is simple to guard against this kind of danger by checking with the pendulum to find out whether or not the particular place you have selected is safe for your meditation.

Meditation can bestow many benefits on us. Properly done, it will improve our health, eliminate fears, complexes, and psychic blocks, and open up to our consciousness new areas of mental experience. It bestows a new outlook on life. Improperly done, it could cause insanity. Therefore, select a good teacher and use the pendulum.

It is interesting that many pendulum operators consider the pendulum to be a form of meditation. It requires a calm, meditative, and relaxed attitude to be a good operator. Using the pendulum will help you to be a better meditator and meditating regularly will help you become a better pendulum operator.

The Pendulum and Telepathy

Webster's dictionary defines telepathy as 'communication from one mind to another otherwise than through the channels of senses'. That telepathy exists is not in question. There are thousands of recorded incidents of telepathic communication. All of us in our own lives have probably been involved in some form of telepathy. Anyone who has been in love, or who has a very close friend, or who owns a dog or cat has seen telepathy in action. We seem to sense what a loved one feels or is thinking without speech; we get sudden and strange urges to call an old friend whom we haven't seen in years, and then find out that the person has been thinking of us and wanted to talk to us. This kind of thing happens everyday. It is not new. What is in question, and what parapsychologists are researching are the fundamental laws governing telepathy. Once these are known man's telepathic powers will be brought under conscious control; we will be able to turn it on or off at will. In short, we will have a new and vastly superior form of communication than anything we now possess. A form of communication that is independent of wires, hookups, machinery, and all external gadgetry. Moreover, we will have a form of communication that does not depend on words or language. It is more direct, precise, and sharp. Theoretically, when perfected and made conscious, telepathic communication will mean the end of all semantic misunderstandings among people.

At the present stage of man's evolution there are relatively few people who are developed to the point of conscious telepaths. The rest of us experience telepathy on a hit or miss basis. Hopefully, more of us will continue to develop this remarkable tool of communication.

I had been using the pendulum for about a year when I accidentally discovered its use as a telepathic device. I was vice-president of a credit consulting firm at the time. Greg Neilsen (co-author of this book) was working with me. My partner, the president of the firm was away on his honeymoon. He and his bride were travelling cross country in a camper and roughing it. Obviously, there was no phone number where he could be reached, but this was the way he wanted it.

He had been gone about a week. I was doing double duty as both administrator and head of public relations and sales. I was buried under this work load, when two burly men walked in and announced that they were I.R.S. agents. They were investigating some transactions we had had with a now-defunct corporation. They wanted to see the records of those transactions for a court case they were preparing against the principals of that company. I was in a quandary. Only my partner had access to those files. I had no idea even where to begin to look. I told this to the agents. They did not want to hear any excuses. They wanted those files and if I couldn't produce them by the next day, they would impound all our books and find them on their own. This did not excite me. I do not know of any businessman who would be particularly pleased at the prospect of an I.R.S. audit, and we definitely could not risk one. In those days our book-keeping was shoddy at best. Although we paid our taxes, there were definitely many minor book-keeping flaws that these experts could have found if they really scrutinized the books.

I had to get in touch with my partner at all costs. But how? Out of desperation I called Greg into my office and told him what was going on. He suggested that we attempt to use the pendulum to contact Steve, my partner. 'Let's hold the pendulum over a picture of Steve, and concentrate on him, and visualize him calling the office. It's our only chance.' I agreed.

We locked the door to the office, held the pendulum over Steve's portrait and began to concentrate. As we concentrated, the pendulum swung rapidly in a clockwise direction, the strength of the swing seemed to reflect the power of our concentration. After about three minutes the pendulum stopped swinging. Both of us felt a mental release even though our concentration was still focused. It seemed as if the pendulum was telling us that the message had gone through – that the resistance had been overcome, and there was no need for further effort. Both of us felt a strange feeling of calm.

Fifteen minutes later Steve called. He was calling from a petrol station near the Grand Canyon. It had taken him fifteen minutes to get there. He said that he had an urge to call the office and find out what was going on, and he seemed to have no idea that we were calling him. He thought that he had originated the idea. I quickly told him what was happening, and he gave me verbal directions where to locate the files, and the whole affair was settled the next day.

I have told the story to many people. Some say that it could have been done without the pendulum. Perhaps, this is so. But Greg and I have tried the experiment a number of times, with and without the pendulum. It always worked better and quicker with it. The only explanation we can offer is that because the pendulum is a physical object, it gives the mind a concrete point on which to focus, thus strengthening the concentration and increasing the mental power available.

Using telepathy without the pendulum seems to require a greater output of mental energy. In any event, we hope that others will experiment along these lines so that new data can be added to this most important New Age science.

10

Exploring the Hidden Powers of Your Mind

Tracking Down Fears, Subconscious Blocks and Complexes

Isidore Friedman, a master pendulum operator, has often said, 'Mankind's greatest danger is having at our disposal the immense powers of modern technology while living in a psychological atmosphere of 2000 B.C. It is tantamount to giving a primitive, nomadic caveman the controls of a 747 jet and telling him to fly it. Only disaster can ensue.'

Most of us are full of outdated, unscientific, and false-to-fact opinions, axioms, and concepts, some of which are conscious but most of them are unconscious. These false axioms or beliefs come from a variety of sources. Some come from antiquated moral or ethical systems, formulated thousands of years ago by pre-scientific people living in a pre-scientific age. Some come from the wrong evaluations that we gave to certain negative experiences in childhood. According to occult science many of these axioms come from negative patterns acquired from past lives. Whatever the source, the problem remains: how do we track these things down and eliminate them so that we can make manifest that which is within us.

Many talented people are held back in life because of inherited feelings of worthlessness or unworthiness left over from childhood incidents. These feelings or emotions are attached to certain beliefs which now have no relevance to the adult state of mind nor to the quality of the actual abilities attained. But the feelings remain buried in the subconscious and still radiate outwardly in the person's aura so that the people whom the person has to impress with his ability, somehow get the feeling that the person is not 'good enough'. On a non-verbal level these subconscious feelings of unworthiness emanate and the person's radiations repel new and better opportunities.

When the person remains unconscious of his radiations and has no understanding of the laws of radiational physics, he tends to duplicate past patterns of rejection, and to interpret them in the same way. When he gets rejected in a job, the old feelings of worthlessness or unworthiness become conscious again, and the person's faith in himself is shaken. This goes on in a negative and vicious cycle which produces hate, anger, frustration, and disease. And the radiations of these negative emotions bring more disaster into the life. And so it goes. The pattern is very familiar.

One of the basic axioms of psychotherapy is that if a person can make these subconscious feelings conscious, he can then deal with them in a rational and ordered way, to dissolve them. Most psychiatrists and therapists attempt to make the person look at their past negative experiences, fears, likes, and dislikes from the viewpoint of the present. The patient sees that there is no point in holding on to old beliefs and resentments in the context of his present life and state of maturity. Those beliefs and fears and resentments were normal for a child but not for an adult; and as the patient looks at these things from his viewpoint, as he rationalizes them – brings reason to them, measures them, they dissolve and discharge their negative energy and the patient is cured.

One of best techniques for diagnosing emotional and mental problems is through using the pendulum. It is a real timesaver. The therapist or patient does not have to spend years of blind probing to find out the root of his problem. With the pendulum he can tune in to and identify any negative energy in his aura and then discharge or neutralize it, as we shall see.

Volney Mathieson was a pioneer in the discovery that all fears, feelings and resentments – all thought and emotion – were electrical in nature. He found through experiments with lie detectors that when a person was reminded of certain past events, or when a change of mood was induced in him, the needle in the detector would jump erratically. The degree of the jumps were in exact proportion to the strength and violence of his reactions.

Mathieson then proceeded to invent the 'electropsychometer' or 'E-meter' which was designed to measure the jumps in emotional and mental current of anyone holding the two handles connected to the electrodes.

He developed a wordlist which was designed to be used in conjunction with the E-meter. He would have the subject hooked in to the E-meter and then have him read off a list of words. Invariably, certain words would trigger violent readings on the E-meter. Whenever this was the case, Mathieson knew that these words were associated with certain very violent and negative fear or resentment complexes in the subject's subconscious mind. Most of the time the subject was completely unaware that he was reacting negatively. His reactions and memories were deeply buried in the subconscious.

This, by the way is a normal phenomenon. It is a basic protective mechanism of the mind. When a person has feelings or experiences of a hurtful nature which are too intense and painful to be borne by the consciousness, the subconscious builds a mental wall around the area. It keeps these feelings and memories hidden 'under the rug', so to speak. If the mind did not build these protective blocks, these memories would flood the consciousness any time something of a similar nature occurred and would probably cause insanity to an untrained mind. So what usually happens is that the memories are triggered, but they are kept below the threshold of consciousness. The person may get violent or depressed but is not consciously aware of the real cause. And here lies real danger.

The person, being unaware of the thrust and violence of his past memories and blocks, projects these energies onto the environment. 'The other guy got me mad.' 'My boss is a tyrant.' 'It's the fault of the Jews, the hippies, the commies, the blacks', and on and on. This is perhaps one of the main causes for failure in life.

Another negative side effect is a horrible waste of energy. Subconscious blocks freeze huge portions of our life energy. They make us tense and tired. Energy which could be used in attaining our heartfelt needs and goals is locked up in the subconscious and wasted.

What Volney Mathieson did was record all the words which produced erratic meter readings, and have the person talk about them. As the subject discussed his associations with the word, the meter reading would gradually lessen to a normal reading. When this happened, it indicated that the locked up energy was released, and that the complex or block was discharged. The process is similar to discharging a battery.

Here is a partial list of the wordlist compiled by Volney Mathieson in his book *Super Visualization*.

> hot — sandy — blue — funeral — hit — fall — germs — deserted — rape — menstrual — shame — scold — raise eyebrows at — cat — rich — lake — woods — flowers — picture show — mean — bridge — snow — cold — white — high — lash — kick — bleed — ache — weep — abortion — love — look down upon — quarrel — to buy — to kiss — to dance — hypocrite — woman — money — water — camping — desire.

For the complete list see the book. We need not get stuck on specific wordlists. It is easy to compile a list of your own. Just choose fifty words at random from the newspaper or your favourite magazine. If you like, you can have a friend choose the words. If you own an E-meter, you can check each other out. If you do not own one, they are expensive, you can use the pendulum as a substitute.

Take the word list and either say the word or point to the word with your finger and think of it and see what happens to the pendulum. If it swings in a negative direction, or if it starts to gyrate wildly and jump around, you know that there is a lot of negative energy being generated in your subconscious from your associations with the word. If the pendulum swings smoothly in the positive direction, it means that the word is not triggering off any violent reactions.

This is an excellent diagnostic tool for professional therapists. It gives them an accurate point of departure for starting treatment.

If you are not seeing a therapist, make a list of all the words which produced erratic pendulum readings. Take a half-hour a day and one by one write out your associations and memories triggered off by the word. You will get tremendous insights into yourself help bring these unconscious blocks to the light of your consciousness. This is the first major step to eliminating them.

CAUTION: If any memory triggers too violent a response, if your reaction is producing strong physiological reactions, then you should

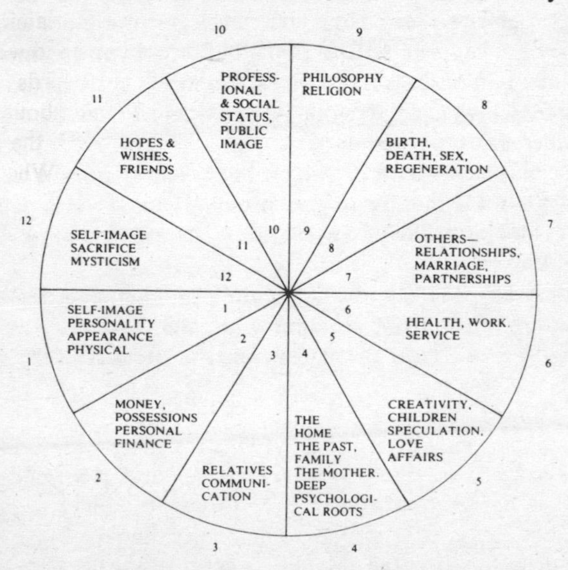

Figure 7. Friedman's 'stick-point analysis' chart.

seek professional help. Do not continue with this procedure. This procedure requires a trained and stable mind. If you do not fall into this category, find someone who does.

Isidore Friedman has a different system for diagnosing subconscious blocks and problems. He calls it the 'Friedman method of Stick-Point Analysis'. It is based on an astrological pattern (see Figure 7). If you make a circle of the zodiacal houses, you will find that the twelve houses together encompass all possible areas of human experience. By holding the pendulum over each of these arcs of experience and noting the kind of reading you get, you will be able to pinpoint exactly which area of experience you are negative in. This area will need work and conscious effort to straighten out.

If, for example, the pendulum reads negative over the third house, it means that you have negative and blocked energies which are causing bad attitudes in the area of your communications with people. You have communication problems. Also you have negative attitudes toward your relatives.

If the pendulum reads negative or oscillates erratically over the fifth arc of experience, it means you have blocks and negative attitudes toward love affairs, creativity, and children. If you investigate these areas honestly, if you go back in your memory and recall all your experiences in these areas, you will find that you probably have always had problems there. Now you can begin to explore them and see why they are happening. Check whether or not you are holding on to a false assumption or subconscious axiom which is negative and false-to-fact and which therefore is drawing to you all kinds of negative experiences for the future.

If you do this sincerely, you will probably find that some negative, painful, past event is modifying your present opinions on a particular subject, and looking at the past event rationally will help bring it under your control.

Check on your progress by holding the pendulum over the area of experience you have been working with and asking, 'Am I still negative in this area?' Note the reading and the intensity of the swing.

Monitoring Your Life Focus

There is a disease in Western society which is more widespread and more dangerous than cancer, heart disease, and tuberculosis put together. It is devastatingly destructive to people's inner health and well-being, and is a greater menace to humanity than the Black Plague was to Europeans in the Middle Ages. We call this disease 'monoideasim' or 'monolaterality'. It is the disease of one-sidedness.

The society is fragmented into legions of specialists who know only one field, and who can not see the relationship of their speciality to the larger whole. The disease is characterized by people's inability to see wholes.

This inability to deal with knowledge wholistically seems to produce people who cannot live and function wholistically. There seems to be little attempt at integrating the various spheres of everyday living into a balanced and symmetrically integrated system. A person may be a successful businessman, but is a failure in his personal relationships. Another may be socially popular, but can not seem to earn a living. Another may be a genius with his hands, but unable to think or plan. Great intellectuals are often out of touch with their feelings, and are emotionally sick. And so it goes. However, it is possible for these people to innoculate themselves against the psychic disease of one-sidedness by consciously attempting to balance their lives.

Isidore Friedman in his book *The Mathematics of Consciousness* outlines these five areas that should be balanced and integrated to live decently. These are the physical aspect (for example, food, clothing, good health, exercise); the mental aspect (for example, intellectual ability, mental interests, the ability to communicate, the sharing of ideas); the financial aspect (career, earning enough money to live comfortably and healthily, making money at what you love doing); the social aspect (friends, emotional nourishment from people of your own kind, love); and the spiritual aspect (the search for meaning and purpose in life, inner growth).

Unhappiness, frustration, and most psychic ills can be traced to an imbalance of function in one or more of these areas. Friedman uses a five-pointed star chart (see Figure 8) to determine which life area is unbalanced and where corrective measures have to be taken. Look at Figure 8 and copy it on a separate piece of paper. Put your name in the centre (or the name of any person whom you happen to be checking). Hold your pendulum over each point in the star and ask: 'Am I unbalanced in this area of my life?' The pendulum will accurately indicate the areas in which you are most unbalanced. Make a list of these. Then ask: 'Does this area need more focus? Less focus? Should this area take priority at this stage of my life at this time?'

Note the answer paying particular attention to the strength and amplitude of the swing. Then begin to work in the areas indicated. For example, if the pendulum indicates that you are deficient socially, begin to make a determined effort to make new friends, to go out, to socialize, and so forth. Perhaps, you are too intent on the financial aspect of your life, you may need the spiritual aspect to balance it.

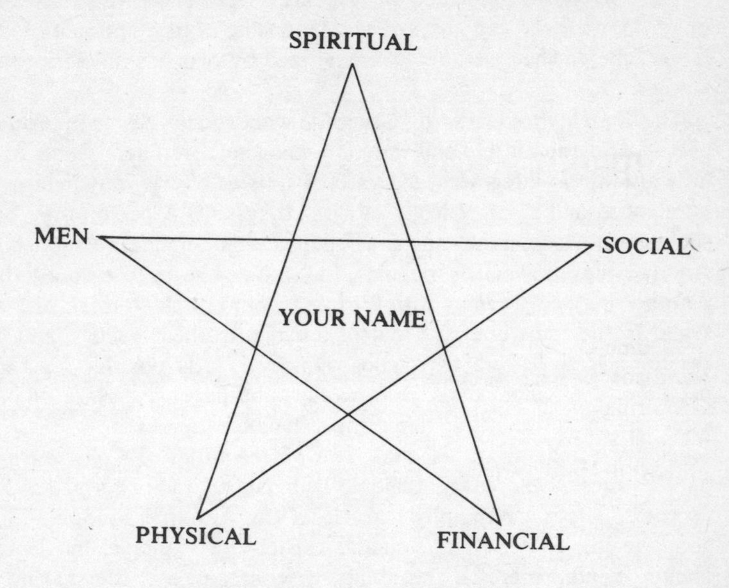

Figure 8. Friedman's five-pointed star chart.

Using this procedure you have a constant monitor of the output of your energies. It is wise to run this kind of a check once a month. If you do these exercises and techniques, persistently and sincerely, the rewards you will accrue in every area of your life will be geometrically greater than the time and effort spent in doing them.

11

The Pendulum and Dynamic Self-healing

Sometime during the early part of this century, Dr Albert Abrams, professor of pathology at Stanford University Medical School in California, was giving a demonstration similar to Gregs to his class of medical students. He chose the healthiest and burliest member, stood him up in front of the class, turned facing magnetic west and began tapping the abdomen. As expected, there were no weird sounds. The percussion method, as this was called, yielded hollow sounds of a normal and healthy person.

Then Dr Abrams brought in a cancer patient, who was in his sixties and in the advanced stage of the disease. The doctor percussed the man's stomach and sure enough, just under the left rib cage in the upper part of the abdomen, there was a dull thud. It sounded as if that portion of the abdomen was filled with solid substance instead of healthy tissue.

Then another patient was brought in. He was a tuberculosis patient. Dr Abrams began percussing this patient's abdomen, and just below the navel, he found the familiar dull thud.

Patient after patient was brought in front of the class. The maladies ranged from cancer to meningitis, from pneumonia to the common cold, but as he percussed each patient's abdomen the results were always the same. Each disease produced a dull sound in specific areas of the patient's abdomen. It seemed that each disease affected a person's nerve reflexes in a unique way.

This discovery alone would have assured Dr Abrams a prominent place in medical history, but he went a step further. Using electrodes attached to a box which contained variable resistors, he wired the sick cancer patient to the healthy medical student. Ignoring the patient, he began tapping the abdomen of the burly twenty year old. As the doctor expected he found the dull thud in the same spot as on the patient. In other words, a disease in one person could, and did register 'electronic reactions' in the healthy body of another. And if he could differentiate between the wavelengths of the various diseases then healthy people could be used to detect diseases in others. The healthy human nervous system could be used as a tool for diagnosis. The science of medical radiesthesia and radionics was officially born.

The pendulum or the radiesthetic method, when properly used, in the hands of a medical expert, is perhaps the most accurate and penetrating diagnostic device known to man. The same faculty which enables a diviner to flinch when near water, and which caused specific nerve reflexes in Dr Abrams' medical student, is now used for detecting disease.

There are numerous and various methods for diagnosing disease through the pendulum. In England, France, and Italy medical doctors have been using the pendulum for years and they consider it part of their normal medical practice. Consequently the literature on the subject is vast, each practitioner espousing his own method. We therefore intend to present the methods which are simplest to use and which can be practiced by doctors on their patients or used by laymen for their own personal health.

Discovering Disease

Make a sketch of the human body or trace the one in Figure 9. It does not have to be artistically exact, just as long as your mind gets the general idea of where the parts of the body are located. The purpose of using this diagram is to assist your concentration, so that your thoughts will be in strong resonance to the person you are checking.

Write the person's name on top of the chart. Many practitioners like to use samples from the patient or 'witnesses', as they are called in medical radionics. You obtain a lock of hair, or sample of urine, or drop of blood, or saliva specimen from the person and place it on the diagram. We have found that a trained mind that can concentrate does not need an actual specimen, but if you find this technique is helpful to you, by all means use it. Hold your pendulum over the diagram and begin to move it over the various portions of the body. You do not have to ask any questions at this point. If the area you are checking is normal, the pendulum will swing smoothly clockwise. You will get a good and smooth feeling all the way up your arm. Keep moving the pendulum over all the areas of the body – move it over the head, the neck, trunk, arms, and legs, and so on. If at any time the pendulum begins to move counterclockwise, or if it oscillates wildly and unevenly, you can be sure that there is unbalanced energy emanating from that area of the body. Something is wrong.

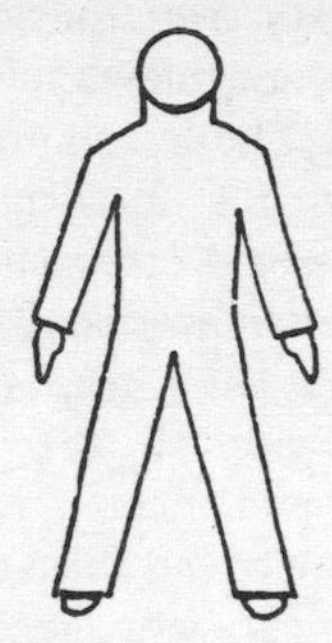

Figure 9. Diagram of human body to assist concentration.

Since the diagram is not too exact, you cannot be certain which specific organ or structure is malfunctioning. You have only identified the general problem area. Next, make a list of all the organs located in that area. If you have access to anatomical charts which have pictures of these organs, so much the better. Now, by a process of elimination, ask the pendulum, 'Is this organ malfunctioning?' and observe the response. Ask the question over each of the organs on your list. Eventually, you will locate the exact organ, after which, you can ask more detailed questions: 'Is the organ inflamed?' 'Is the organ hyperactive?' 'Is it hypoactive?' The pendulum will give you the answers if you know what to ask it. This is why a solid knowledge of anatomy and physiology is very useful here. It will help you to know the appropriate questions to ask.

One of the beautiful aspects of this method of diagnosing is that the subject does not have to be in your presence. All you need is the name and your home-made diagrams, or anatomy charts. Of course, if the person is present, all you would have to do is hold the pendulum over each part of his body. You could dispense with charts. But the basic procedure would be the same.

Whenever you are diagnosing at a distance, or (as we shall see later) broadcasting remedies from a distance, you are using the technique of *teleradiesthesia*. It operates on the same principles as TV or radio waves, with the operator's mind as the transmitting and receiving instrument.

Diagnosing the Subtle Bodies

One of the reasons for the abysmal failure of modern medicine and the growth of alternate healing systems, such as herbs, chiropractic, homoeopathy, acupuncture, and reflexology is that the premises on which it is based are completely false-to-fact.

Most modern practitioners treat the human being as a physical body, and more often than not, they alleviate the symptoms without ever treating the true cause. Those who do take into account psychological factors, still make the fatal mistake of separating and segmenting the multidimensional organism-as-a-whole which we call the human being. They see man as a body AND a mind; or as a body AND a soul, rather than as a mental-emotional-physical being functioning as a unity, with all the parts interconnected and interreacting, each affecting and affected by the other.

Consider the consequences of this false-to-fact view of man. If he is nothing more than a physical body, then it is perfectly logical to correct his malfunctions in the same way you would correct any other machine or mechanical instrument. If a part malfunctions, just cut it out and replace it with another. Barbaric things are done to the physical body without any knowledge of how they affect the rest of the organism-as-a-whole. What is worse are the crude diagnostic and testing procedures and devices practitioners use. A friend of mine almost died in the hospital, not from his malady which was a relatively minor hernia operation, but from the dozens of tests that were performed on him. He is a sensitive person with a highly refined nervous system. Disregarding this fact – the doctors were completely unaware of it – they injected him with barium enemas, bombarded him with dangerous X-rays, and kept him in a large noisy ward where it was impossible for him to get any rest. This kind of thing happens all the time.

With a simple tool like the pendulum, a trained practitioner who is willing to take the time to master its secrets, can diagnose anything accurately and safely with no ill side effects to the patient. In Europe, pendulum diagnosing is done everyday. The mere hint that there is a better method should spur those in control of the country's health care to spend a few million pounds researching it. We spend billions on new and more efficient machines that rape the organism but very little on the human system which is the most sensitive and accurate detecting instrument known to us at this time.

That modern medicine treats symptoms and ignores causes is such an oft repeated criticism that it has become a cliché. Unfortunately, it is quite true. However, given the premises on which most health professionals function, it is inevitable. How can they get at causes when they completely ignore the higher octaves of man's structure – his finer bodies?

The existence of man's subtle or finer bodies has been known for millennia. Seers, ancient and modern, have testified to their existence and described them in great detail. Latter day radiesthetic and radionic research has confirmed these facts. Further confirmation has come from the latest techniques in high-voltage photography (Kirlian photography), in which etheric, astral, and mental phenomena have actually been photographed.

Besides the physical body, which is nothing more than a vehicle, man's subtle anatomy consists of an etheric body, which is sometimes referred to as the 'energy body', an astral or emotional body, and a mental body.

The true being of man lies beyond these bodies. These bodies are merely vehicles through which the 'power-to-be-conscious' or the 'innate intelligence' operates on this dimension of existence. The *physical body* is the vehicle through which the power-to-be-conscious expresses itself in action. The *etheric body* is the vehicle through which the power-to-be-conscious energizes and animates the physical body. It is to the physical body what the plumbing and electrical wiring are to a house. It supplies the physical with vitality and energy. The *emotional body* is the vehicle through which the power-to-be-conscious feels on the emotional plane. It is that part of us that can express and respond to emotions. The *mental body* is the vehicle with which the power-to-be-conscious thinks.

When a person is truly healthy, all these bodies are active, poised, and in harmony with each other. Most diseases began in the inner bodies and work their way down the dimensions. The process is akin to vapour becoming water and then ice. A gradual solidification and hardening of the diseased vibration takes place until it finally manifests in the physical body. Thus what most modern physicians concern themselves with, when they treat the sick, are really the end-products of a long illness-producing process from the inner levels.

Dr George Goodheart of Detroit has shown in his kinesiology research that vitality enters the body from the outside rather than from within. First, he tested a strong muscle and then placed it behind a lead shield so that no outside radiations could reach it, and then he retested the muscle. Invariably, it was weaker than with the first test.

Some modern-day theorists maintain that man is kept alive by a ray from the cosmos. If the energy from this ray is blocked in any way, then to that degree the vitality and strength of the organism diminishes. This diminution in energy is the beginning of all disease.

It would seem safe to assert that all disease, whether it is in the physical or one of the finer bodies, is caused by some block of the free flow of the natural life energy. Any block in the finer bodies will prevent the cosmic energies from entering into us properly. This is why it is of the utmost importance, when treating a family member or oneself, to know where the block is rooted. There is no question that whenever a person is sick, a block exists somewhere within his being. The procedure for locating the block is as follows:

Take a piece of paper and write the person's name on top. If you prefer to use a sample or specimen from the person, all well and good. Underneath the name write the words:

PHYSICAL BODY
ETHERIC BODY
ASTRAL BODY
MENTAL BODY

Hold the pendulum over the paper and while pointing with your left hand to the words PHYSICAL BODY (If you are left-handed you would point with your right hand), ask 'Is John Doe's physical body healthy?' If the pendulum swings positive, the physical body is healthy; if it swings negative, there is a problem there. Write your positive or negative findings next to the word. Next, repeat the procedure over the word ETHERIC BODY and again record your results. Then repeat it over the words ASTRAL and MENTAL BODIES, each time recording your results next to the words. If only one body of the four is negative, the person is in relatively good health. Two out of four is so-so. Three out of four negatives is serious and usually means the person needs a lot of rest. Four negatives out of four is very dangerous – usually an emergency situation. If such a reading persists for a long period of time, the patient is probably headed for a complete breakdown.

Each of the bodies has different laws of treatment. You treat the physical body through physical means: diet, exercise, massage, and medicine when required. If there is actual structural damage, surgery may be indicated.

The etheric body should be treated by rhythmic breathing and rhythmic motions. Fresh air and sunshine, and baths are also helpful. When either physical or etheric body is negative, make sure the person gets plenty of rest. He may be suffering from fatigue which is common in our hectic society.

We treat the emotional body through the arts. Creative activity, warm social relationships with harmonious people, colour therapy, and good music are all excellent therapy.

The mental body is more difficult to treat. Sometimes it reads negative, temporarily, caused by too much thinking which produces mental fatigue. When the mental body continually reads negative over a long period of time, it is necessary to probe the person's thought patterns to see why he is thinking negative thoughts. The person would have to make an effort to develop new mental interests to stimulate a more positive mental outlook. Or he may have to learn to turn his mind off periodically to give it a rest. Meditation and concentration exercises would be helpful.

By learning to diagnose and treat the finer bodies, health professionals could raise their patients recovery rate, significantly. If, for example, they were treating a physician ailment, but found that the cause lay in one of the other subtle bodies, they could continue to treat the physical symptoms and relieve any discomfort the patient was having, but in addition they would prescribe the appropriate remedies for the other body or bodies. Thus, they would effect a more permanent cure.

Choosing a Doctor

If through ignorance of the laws of life and through wrong habits of living we become diseased (un-eased, unbalanced), and the disease does not respond to our own efforts of cure, the next important step in the healing process would be to choose the best doctor or healer for you. As in any profession, there are proficient and inefficient doctors. And even among the qualified ones, there are some who would be better for you than others. It all depends on their vibrational harmony with you, at a given time. Most successful doctors, though they probably do not realize it, do more for their patients by the strength and stability of their auras than with the remedies they prescribe. Both are necessary. To make sure you get the benefit of both factors and thus increase your chances of a solid cure, use the pendulum to select your doctor.

Follow the procedure mentioned in other parts of the book. Make a list of the possibilities and hold the pendulum over each of their names. Ask, 'Is this the right doctor for me at this time?'

As a further test, write your name next to the name of the doctor whom you are thinking of selecting and hold the pendulum over both names, and ask, 'Are we compatible?' Check your answer. It is not wise to see a doctor who reads incompatible with you no matter how many of your friends rave about him. Remember we are dealing with dynamic energies, not logic.

Therapies – Treating Disease

Since the pendulum through the medium of the human nervous system is an excellent measuring tool, its use is not limited to diagnostics. It is also excellent for all forms of therapy where dosage, frequency, and remedy selection are necessary.

Diet

According to Dr Alfred Pecora, a noted nutritionist, most diseases are caused by improper dietary habits. Most people are ignorant of the basic dietary laws of the human organism. Even worse, the role of nutrition in health is abysmally ignored or underplayed by the medical profession and by the western way of life.

We seem to be more careful and scientific about feeding our livestock than we are about feeding ourselves. Also many people take better care in selecting the proper fuel for machines, cars, jets and boats than they do in feeding their children. Not only could most diseases be prevented by correct dietary habits but also many of our major discomforts could actually be cured by switching to a healthy unadulterated diet that is tailored to our body's needs.

According to Dr Pecora, arthritis is a prime example. Medical science calls it 'incurable', yet there are many recorded arthritis cures that have resulted from a change of diet.

Also it is one of the most preventable diseases if a person would eat as many raw foods as possible, and eliminate white sugar and white flour from his diet.

The problem with diet, as with so many other areas of life, is that there are few general rules that are suitable for everybody. It seems that the right diet for a person is as unique to him as his belt size and his fingerprints. What works for one may be total poison for another. This is why it is not wise to blindly follow any miracle diet that claims to do wonders for everybody. We are all unique, and so are our dietary needs.

The pendulum gives each and everyone of us a tool for creating our own unique and specific dietary patterns. You do not have to read numerous books or become a nutrition expert. All you do is hold the pendulum over any piece of food that you want to eat, ask, 'Is this good for me now?' The pendulum will give you an exact reading of the effects of that particular food at that particular time.

And here we will share some interesting facts about food which are not included in most books on nutrition. A food may be beneficial for you generally, but bad for you at a specific time because of your mood, energy level, or state of health. A food may normally be bad for you but at a given time it may be just the thing you need to fill some deficiency that your body needs.

Without the pendulum we lose the dynamics of food selection. We are forced to eat by theories and general principles rather than by what is uniquely right for our bodily structures and way of life.

Dr Arnold Forster brought to my attention another interesting and important factor about diet. Two foods may be good for you, but eating them in combination could be harmful as the energy of one may offset the energy of the other. Or, as with acids and starches, one food may delay or block the digestion of the other, thus causing stagnation in the intestines.

Food combinations are another new area of inquiry, but with the pendulum they can be checked safely and accurately. Hold the pendulum over any combination of foods and ask, 'Are these two foods compatible?' The procedure is the same as if you were running an ordinary compatibility analysis.

The results of using your pendulum to monitor your bodily fuel intake could save you hundreds of dollars in doctor bills and add years to your life.

Vitamins

About five years ago, I was under a lot of heavy stress because of the sickness and death of a close family member. The sickness had lasted two years and when it was all over, I, myself, became ill. My family doctor ran all kinds of tests on me and after checking my blood, he told me that I had mononucleosis, and would have to go to the hospital for six to eight weeks.

After having spent two years in and out of different hospitals, I was in no mood to go back. I did not ever want to see one of those places again. I almost preferred taking my chances at home even if it meant delaying my recovery.

When I told my problem to a friend, he introduced me to a pendulumist, an expert. The man spoke to me over the phone and told me that my problem was a severe vitamin deficiency. He suggested a long list of vitamins to take and even indicated the amount and the frequency of intake. I followed his advice, and three days later I had regained my full strength. When I went to my doctor, he checked my blood again. When the blood analysis came back from the lab two days later, he scratched his head and said that all traces of the disease had completely left my blood, and that it was now completely normal. He did not understand it, but he said it was all right for me to pursue my normal lifestyle.

This was my introduction to the beneficial effects of vitamins. I have been taking them regularly since then, and have yet to experience any severe health problem. I had always been prone to getting colds and sore throats, but these have now drastically diminished both in severity and frequency. I know many other people who have had similar experiences.

One of the problems with vitamins is determining the exact kind and amount of vitamins and frequency of intake. Since each person is continually changing, the stresses on him are also changing, his energy levels are changing; the delicate chemical balances within the organism are continually changing, so it naturally follows that his vitamin needs will also be constantly changing. It is not possible to monitor these changes from a book that gives you just standard recommended dosages.

Also do not be deceived by the 'minimum daily requirement' dosages published by the government. Through radiesthetic analysis, we have found these to be pitifully inadequate amounts for many people's needs.

If you are feeling well, you can maintain that feeling by checking every week or month on your vitamin needs. Make a list of the major vitamins, such as Vitamins A, C, D, E, and the varieties of B vitamins, as well as minerals, and enzymes.

Just ask the pendulum whether you need a particular vitamin, and then ask what dosage is necessary. For example, if the pendulum indicates that you need vitamin E, ask 'Should I take 100 International Units a day?' If you need more, the pendulum will swing positive; if you need less, the pendulum will swing negative. If the pendulum swings negative, ask your question with a lower figure, until you get a positive reading. That is your dosage.

If the pendulum reads positive on 100 I.U., continue upping the numbers of units in your question until the pendulum either oscillates or swings negative. That is your dosage. This process is almost like reading an electric meter. The pendulum will swing in one direction until you identify the correct dosage.

Many vitamin manufacturers put out combination formulas or multivitamin capsules. Some are good, some are not. If you do not feel like taking many separate pills and wish to take one capsule that has everything you require, check out the formula with the pendulum.

Once you know the dosage, or how much of a vitamin, you need per day, then you can ask whether you should take it all at once, or twice a day, or three times a day. Just ask one question at a time, phrased so that a yes or no answer can be given. By a process of elimination you'll arrive at your answer.

Very often when a person is sick, the pendulum will indicate very high dosages of vitamins. Do not be alarmed for this seems to be normal. I have seen people for whom the pendulum indicated 2000 to 3000 I.U. of vitamin E a day, and 10,000 to 15,000 mg. of vitamin C a day. The body obviously needs tremendous amounts of certain substances so that it can carry on the healing process, and also counteract the effects of air pollution on our bodies.

The important thing is to develop enough skill with your pendulum so that you trust it implicitly. If, in a health matter, you have doubts about your readings, it would be wise to have another pendulum operator, who is neutral, check your findings.

Cell Salts

Cellular therapy, or *Biochemic System of Medicine*, is the name of a system of treatment developed by Dr Wilhelm Heinrich Schuessler in the late nineteenth century, based on the 'cell theory' of Virchow. In 1858, Virchow discovered that the body is little more than a huge chemical plant and that what we call disease is nothing more than a deficiency in one or more of the inorganic chemical constituents in the body's cells, he found that all cells are composed of three basic elements: water, organic matter (such as sugar, proteins, and fats), and inorganic materials, that is, mineral salts. These 'cell salts' or mineral salts, or tissue salts are essential in the rebuilding of cells. These salts occur in minute quantities throughout the body, but their role in fighting disease and balancing the body is immense.

By diagnosing which tissue or cell salts a patient was deficient in and by correcting the deficiency by giving the particular cell salt to the patient in the form of tiny pills prepared homeopathically, Dr Schuessler compiled an incredible record of cures. The beauty of his system lay in its simplicity. There are twelve basic tissue salts and by using one or combinations of them, most diseases can be cured.

These tissue salts are not drugs, and they are harmless to the system. They are merely minute doses of minerals that act to restore bodily balance. They are not designed to suppress symptoms or kill bacteria. When the balance is restored, the body will proceed to heal itself. The twelve cell salts are as follows:

1. Calcium Fluoride
2. Calcium Phosphate
3. Calcium Sulphur
4. Ferrum Phosphate or Phosphate of Iron
5. Kali Muriaticum or Potassium Chloride
6. Kali Phosphoricum or Potassium Phosphate

7. Kali Sulphuricum or Potassium Sulphate
8. Magnesium Phosphoricum or Magnesium Phosphate
9. Natrum Muriaticum or Sodium Chloride
10. Natrum Phosphoricum or Sodium Phosphate
11. Natrum Sulphuricum or Sodium Sulphate
12. Slicea or Silicic Oxide

These salts come in various potencies from 1x to 12x. The higher the number, the more it has been potentized and the higher the potency or frequency. These tissue salts are directly assimilated into the blood stream and act immediately. (For detailed description read *The Biochemic Handbook.*)

An illustration of this, I watched my friend take some silicea for a boil. Within minutes we heard a sharp crack on his arm where the boil was. There was an intensification of the pain – he described it as an explosion of pain – and then the size of the boil gradually diminished. It took an hour and a half for it to go down halfway in size. Within four hours, the boil was gone, and he had no more pain.

As with vitamins the problem with these miracle salts is determining which salts to take for what condition, the dosages, and frequency of intake. Follow the same procedure with the pendulum as with the vitamins, and you will have access to a truly safe and effective method of treatment hitherto not available to you.

Locating Spinal Subluxations

A well-adjusted healthy spine is a must for proper functioning because the structure of the spine protects the spinal cord which is one of the main pathways for nerve impulses to travel from the brain to all parts of the body. When a vertebra is out of line, it exerts pressure on the spinal cord and other adjacent vertebrae and blocks the flow of nerve energy needed to maintain physical and muscular coordination. Different vertebrae contain various nerve reflexes which feed different parts and organs of the body.

When a vertebra is out of alignment, we say that a spinal subluxation cxists. This is not serious unless the vertebra continues to be malaligned in this way for a long period of time, or if some structural weakness is causing the subluxation. Subluxations are easily corrected as anyone who consults a chiropractor knows.

There are various methods for locating spinal subluxations. One of the most popular among chiropracters is the analagraph machine. This utilizes a thermocouple device which is passed over the skin all the way down the length of the spine. As the thermocouple passes each vertebra, it picks up any excessive heat readings and records them on a thin sheet of graph paper in the machine. All the doctor has to do is look at the graph paper and note any strong heat peaks. These are the areas where a subluxation is most likely to be present.

Analagraph machines, however, are quite expensive. Using the pendulum, a doctor could get just as accurate a diagnosis at a fraction of the cost, merely by holding the pendulum over each segment of the spine. The pendulum will gyrate clockwise if that area is normal, but as soon as it is passed over a subluxation it will either oscillate or gyrate counterclockwise.

It is also possible to detect spinal subluxations without the presence of the person. Just write the name of the person over a chart of the spine or think of the person's name and the pendulum will swing clockwise or counterclockwise over the chart as accurately as if you were holding it over the actual person. The layman can ask his pendulum if there is a spinal subluxation and whether it is severe enough to warrant an adjustment. Some subluxations are only temporary and after a period of rest adjust themselves.

Colour Therapy

Colour therapy, or chromotherapy as it is often called, is one of the most exciting fields of research in modern therapeutics. Some of the recent workers in this field are Dr Max Luscher, Christopher Hills, George De La Warr, Rudolph Steiner, and Isidore Friedman. Of course, there is really nothing new about it at all.

The therapeutic uses of colour have been known for thousands of years, but the knowledge was confined mostly to esotericists and artists. Some of this knowledge has even filtered down into modern usage. We talk about feeling 'blue', or being 'green' with envy, or being wrapped in a 'brown' study, and so on. It has remained, however, for modern science to give us the scientific basis and the empirical evidence for these intuitive concepts.

Colour is an energy with a definite wavelength and frequency. And this energy can affect the human organism. Many nervous diseases have been permanently cured merely by having the patient change the decor of his bedroom.

Colours can induce peace or stimulate to activity. They can evoke a mood of confidence or distrust; they can stimulate mental activity or cause emotional upheavals. These facts are being used by behavioural psychologists in the colour schemes of banks, offices, and factories.

And besides all this, colours can also heal. The first step in using colour therapy is to find your basic or resonant colour. Make a chart like this, and ink in each colour, if possible:

Your name/red	Your name/blue
Your name/orange	Your name/indigo
Your name/yellow	Your name/violet
Your name/green	

Now make the statement to your subconscious that you are looking for your basic colour. Then hold the pendulum over your name and each colour, in turn. Your resonant colour will be the one that produces the strongest, and smoothest positive swing.

You can learn what each colour means psychologically or therapeutically in the many excellent works on this subject, most notably, *The Luscher Colour Test*, by Dr Max Luscher; *Colour Healing* by Mary Anderson; *Nuclear Evolution* by Christopher Hills.

Once you have determined your resonant colour, begin to wear as much of it as you can, moderated by good taste, of course. You will find that you have more energy and are less prone to sickness. Whenever you feel tired, hold some velvet ribbons of that colour and stare at them. Paint your bedroom in that colour, so that you are bathed in its rays while you sleep.

Our auras, of course, are composed of all the colours in the spectrum in different proportions. A delicate colour balance is maintained in the astral body, and when this balance is disturbed, we have the beginnings of disease. Whenever you feel fatigued or sick or just moody and depressed, it is a good idea to check for colour deficiencies in your aura. Make a chart like so:

Red	Blue
Orange	Indigo
Yellow	Violet
Green	

Hold the pendulum over each colour and ask, 'Am I deficient in red? Am I deficient in orange?' and go on down the list. If the pendulum swings positive over any of the colours, put a check mark next to the colour, and continue your questioning.

When you have finished, you will probably find that you are deficient in two or three colours. This seems to be usual. In rare instances, especially after a violent emotional upheaval, people need more than three, sometimes the whole spectrum. Whatever colours you find yourself deficient in you must restore to yourself. How you can do this varies. Some practitioners use colour lamps, and have the person sit under a certain colour for a specified period of time. If you use this method, you can use the pendulum to check the amount of time. Other practitioners use colour-charged water, which the deficient person drinks. To charge a glass of water, place a colour filter on top of it, and let it stay out in the sun for a certain length of time. The energy from the sun going through the coloured filter gives the water a powerful charge of that colour. You can check the amount of charging time necessary by testing with the pendulum, and also test the amount of water you should drink at a time.

Another and easier way of pumping colours into yourself is through thought. When you think strongly of a thing – anything – you begin to resonate or tune in to it. This is the method we have found to be the easiest. It can be done quickly, wherever you happen to be and you do not need expensive equipment. Let us say, you find yourself deficient in red. Hold the pendulum over your left hand (if you are right-handed, over the right hand if you are left-handed) and visualize the red colour streaming into your hand and from there into the rest of your body. As you think this, you will note that the pendulum will start to swing in a positive direction, very powerfully. The strength of the swing will be in direct proportion to the power and intensity of your thought. If the pendulum is gyrating weakly or lazily, then your concentration is weak and it will take you longer to get the necessary charge. Perhaps, you should rest and try again later when you feel stronger; or you can have someone else do it with you. This way you get the power of his or her thought combined with yours. If the pendulum gyrates strongly, it means that the colour is going into your system properly with little resistance. Keep on visualizing until the pendulum stops swinging. When this happens, it means that you have absorbed as much of the colour as your body could safely take at this time. Repeat this process for every colour you are deficient in. Afterwards you can check your efficiency by re-running the colour check. Ask 'Do I still need red? orange?' and so on.

If you find it difficult to visualize colours, you can look at a coloured ribbon while doing your visualization. This will help you to get the colour clear in your mind.

In the treatment of any disease, colour should play a part in the total treatment. It will strengthen whatever else you are doing to the system in a beneficial way.

Generally speaking, hot colours such as red, orange, and yellow will stimulate and energize. If a person is lethargic and hypoactive, he or she will generally need a good dose of one of the warm colours. Cool colours such as blue, indigo, and violet slow you down and are a good antidote to hyperactivity. They are also good for shrinking inflammations and lowering fevers.

I have seen many a sore throat cured with the cooler colours. I also saw a prostate condition cured by the use of colour alone. A friend was in the hospital undergoing tests. The urologists told him that his prostate gland was enlarged and that it was blocking his urinary flow – a common ailment in older men. The standard medical procedure in such instances is surgery, as this is not considered a major operation. We checked his colour needs with the pendulum, and found that he needed blue, indigo, and violet to shrink the prostate to normal size. Twice a day, for three days, we broadcast these colours to him in his hospital bed. We were at home. Besides the broadcasts he focused his thought on these colours as much as possible. After the three days, the urologist came to make some final checks and found that the prostate condition was back to normal. My friend was discharged a few days later.

These instances of complete cure by colour alone are rare. But as we mentioned, it is a powerful adjunct when used in combination with other therapies.

Human Healing Circuits

For thousands of years, there has been a mystique about 'healers' who could heal by direct contact or by 'the laying on of hands'. In light of what science has uncovered about the nature of the universe and the human body, the procedure is not mysterious at all, and is, in fact, open to most people. But we must learn the ground rules first.

The human body is a kind of intricate electrochemical-mechanical machine. It is also a tetrapolar magnet. That is, a magnet with four poles.

According to L.E. Eeman, one of the great pioneers in this field, in right-handed people the right side of the body is positive and the left side is negative. The head is positive to the feet, and the feet are negative to the head. In left-handed people the body polarity is reversed. The left side is positive and the right negative. The head is negative and the feet positive. See Figure 10.

When one person links up to another with his positive connected to the other person's negative and his negative to the other person's positive, a circuit is established and energy will flow between the two people. It is just like an electrical circuit. When a person is sick, fatigued, or diseased, the condition begins to manifest by a drastic drop in body voltage. A person who maintains his physical voltage at high levels all the time will be immune to all diseases. His aura, or electromagnetic forcefield will protect him. When the body voltage

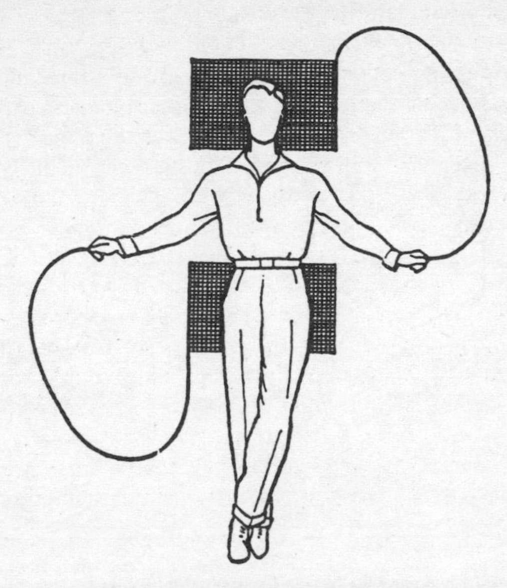

Figure 10. Eeman's energy-flow circuit.

drops, the aura – which acts like a psychic skin – weakens and leaves the organism vulnerable to outside negative influences.

The first step in treating disease is to rebuild the energy level of the person back to normal levels. Strengthen the aura, and the body will take care of itself.

What the so-called magnetic or psychic healer does is create a circuit between himself and the patient and allow his superabundance of vitality to recharge the patient. It is similar to recharging your car battery when the voltage weakens.

The healer will be successful in proportion to the strength of his own energies, and the length of the treatment and the creation of the proper circuit. These are facts that can now be proved, empirically. In Kirlian photography actual photos can be taken of a person in a normal state, in which you can see a bright, healthy aura. But a Kirlian picture of that person when he is sick will show a weak, dull-looking aura, and the Kirlian photo taken of the person after he has been in contact with a psychic healer will show a bright, healthy aura, again.

If you want to treat a sick person in your family, check first to see whether they are right- or left-handed. If both of you are right-handed, sit across from each other and you hold the other person's left hand

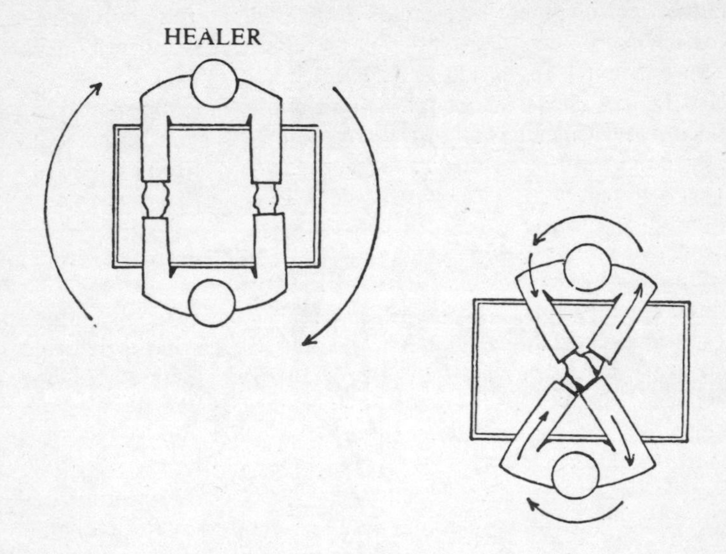

Figure 11. Position for magnetic healing circuit.

with your right hand, and his right hand with your left hand, as shown in Figure 11 (top).

If the person is left-handed and you are right-handed, hold their right hand with your right hand and their left hand with your left as shown in Figure 11 (bottom). Remember the rule is positive goes to negative. CAUTION: DO NOT REVERSE THE POLARITIES. DO NOT HOOK UP POSITIVE TO POSITIVE OR NEGATIVE TO NEGATIVE. THIS CAN BE EXTREMELY DANGEROUS FOR BOTH OF YOU, ESPECIALLY IF THE OTHER PERSON IS VERY WEAK.

One of the problems with magnetic healing is determining the length of the treatment. At any given time there is one optimum length of time during which the most benefit will be derived by both the receiver and the sender. This time can be determined by the pendulum. Also check the frequency of treatment with the pendulum. Use the techniques described previously.

Eeman, in his master work *Cooperative Healing*, describes situations in which with some people his circuits seem to work miraculously and with some other people, it did not work, moreover strange and intense antipathies developed between the patient and himself. After a few minutes in circuit together, they both began to hate the sight of each other, and as Eeman tells it, this strange hate lasted for many years. Not only that but after the treatment both he and the patient felt completely devitalized.

Had Eeman known about radiesthesia at the time, he could have avoided these incidents. And so can you. Before going into circuit with anyone check your compatibility. If your magnetic force-fields are too incongruent and out of harmony with each other, find someone else to administer the treatment.

Eeman also pioneered a variation on the human energy circuit. Believing that bioenergies followed electrical laws, he devised a circuit made of copper mats and wires which could take a person's internal energies and recycle them. One person could use it alone or he could be hooked up in circuit with others. At one time during the experiments, Eeman had as many as twelve people hooked up together. For more information on how this was done read *Cooperative Healing*.

Copper Eeman relaxation circuits are very powerful. One mat goes under the head and the wire leading from it is grasped by the left hand. The second mat goes under the base of the spine in the sacral area, and the wire leading from it is grasped by the right hand.

Eeman had impressive success with this circuit. In his thirty years of practice, many thousands were helped with all kinds of ailments. Again the problem here is the length of time for each treatment. If you have no accurate gauge for monitoring the length of the treatment, do not even attempt to use this circuit. Overdoses can be dangerous.

With the pendulum though we have an accurate gauge, and as with the other therapies, we do not have to go by experience, or trial and error. New possibilities are open for the relief of sickness and pain.

Summary

It is not our intention here to give a complete outline of the entire health field. This would require many volumes. There are many therapies that we have not included such as reflexology, acupuncture and acupressure, hydrotherapy, heat therapy, Bach Flower Remedies, herbal medicine, homoeopathy, and fasting. There are many more that will be developed in the future. And as beautiful as all these therapies are, it is important to remember the essentials of health. These are central. The paths to them are varied.

In this book *Organics: The Law of the Breathing Spiral,* Friedman defines health as 'wholeness'. This means living fully and rhythmically with every part of our being. In ancient times, this concept of wholeness was called 'holiness', and is the real root meaning of the word. Anyone who lives in only part of his being becomes unwhole or unholy. In short, unhealthy.

Learning to live 'wholefully' is without a doubt a long and arduous process. There is so much implied by the statement. One must learn the nature of one's being, and the nature and laws of the vehicles through which it functions. One must learn the laws of the natural order process we call the universe. And finally, we must learn and experience our connection to this process and to see ourselves as mini-processes integrated in ever larger and larger maxi-processes. Man is not only an epitome of the cosmic process but also *is* the cosmic process in miniature.

True health comes when man sees and knows that his every movement and gesture are not personal movements but, in essence, universal movements: that the universe expresses itself through him.

This, of course, is not accomplished overnight. But in our own little way, each according to his understanding and ability can begin to approach this ideal by tiny degrees. And the mere act of starting and growing will automatically make us healthier.

But until we align ourselves to the laws of the natural order (the laws of what is) and become attuned and resonant to them, there will always be ill health, and there will always be the need for newer and better therapies. And perhaps, this is the lesson that every disease manifestation is meant to teach us. It is nature using the rod of pain and discomfort on us to nudge us closer to better resonance to her.

As far as therapies are concerned, everything in this dynamic, ever-shifting and ever-changing universe at one point in space-time is a blessing for us and a health bringer, but conversely at another point in space-time the same thing can be utterly destructive. Some things that are helpful for certain people are poison for others. Even too much of a good thing could be harmful. We need a tool of measurement that registers all these qualities. And we have it in the human nervous system, the radiesthetic sense.

Learn the techniques of radiesthesia. Take the time and the effo[illegible] become reasonably proficient with them, and you will have tak[illegible] giant step to 'wholeness'. You will be living and using another part of your hitherto latent sensory apparatus, and you will become free of the domination of any and all authorities and experts. You will not need them, for you will be a measurer. (In the Hindu philosophies the word for man is *manas* which literally translated means 'the measurer'.)

As you learn more about natural healing techniques, you will find through experience that one therapy rarely yields permanent results. Most of the time, it is a synthesis, a blend of various physical, emotional mental, and spiritual therapies that produce the benefit. This is logical when you realize that most diseases are not caused by one negative factor but by a synthesis of negative factors which gradually add up and take their toll of the body.

So the secret in healing is to find the proper *combination* of therapies which will produce the right resonant chord or frequency which will bring the energies back into equilibrium.

Two therapies when viewed separately or used alone may be excellent but when used in combination may be hazardous. The reverse is also true. Two separate therapies, when viewed and administered separately may not be good for you, but when taken in combination may interact in such a way as to give you exactly the beneficial results you need.

Each disease symptom is unique. Each person is unique. At one time a cold can be cured by acupressure and herbs, whereas at another time the cold will only respond to a dietary change and cell salts. These are variables. With one exception, there are no set rules in the dynamic energy universe where everything is in a constant state of flux and change. The only set rule we can make is 'Discriminate and Measure'.

There are four basic pendulum techniques that should be mastered before engaging in healing work:

The Compatibility Check: This measures the interaction of two or more radiating fields. Since everything in the universe radiates, everything has a force-field and therefore you can check everything for compatibility. Whatever it is you are checking put the two things or people together, either physically or symbolically.

Hold the pendulum over the substances or symbols or photographs and ask, 'Are these compatible with each other?' Watch closely the kind of swing you get and the intensity of it. The Compatibility Check technique is especially important when combining various therapies or foods or herb extracts.

The Simple Yes-No Question: This is the most versatile and simplest pendulum technique to learn and use. All other methods are merely variations on this. In an emergency, it can be used for almost anything. Simply ask any question that can be answered yes or no. By a process of elimination you can obtain any kind of information.

Diagnosing from Diagrams or from the Body: Here you do not have to ask any questions. Just hold the pendulum over each area of the chart or the body and see what kind of movement you get from the pendulum. If it swings positive, the area is alright, if it swings negative, there is something wrong.

This is quicker than asking a question over each organ. Asking that many questions can be quite time consuming and exhausting. Here you are just reading the force-field direct.

Measuring Degrees: This can be applied to measuring how much of a substance to take or measuring how long a therapy treatment should last, or how often a given therapy should be repeated per day, or how long the treatments should be continued.

Whatever the unit of measurement you are using, whether quantities such as milligrams of a vitamin, or dosages of 2x to 12x of a cell salt, or units of time – in minutes, hours, days or weeks, frame your question with that measurement in mind. Start your questioning with the smallest possible quantity and work your way up, all the time carefully watching the movements of the pendulum. At the mention of one quantity, the pendulum will either swing negative (counter-clockwise) or it will oscillate. In any event when the pendulum stops its positive (clockwise) gyrations, that is the true measurement, you are seeking.

For example, you have a cold and the pendulum reads that vitamin C would be helpful. Then you would ask, 'Do I need 100mg? 200mg? 300mg? 400mg? per day? and so on as long as the pendulum keeps moving positive, keep increasing the milligram amount. Let it swing a few times before going on to the next figure. Let us say that the pendulum begins to oscillate when you mention 3000mg a day. Now you want to know how long to keep taking this dosage. So you ask, 'Do I continue this treatment for one day? two days? three days?' and so on. Again wait a few moments after mentioning each figure to give the pendulum time to react to it. Let us say the pendulum stops gyrating when you mention five days. It means that you should take 3000mg of vitamin C every day for five days.

As you get proficient, you will no doubt develop your own methods. This is normal and healthy. In all probability there are no two pendulum operators who work the same way. The whole thing is unique and individualistic. The important thing is to learn the basic principles behind it.

This is a relatively new science and new methods and new research are necessary and welcome. At the end of this volume you will find a comprehensive Bibliography.

12

Become the Master of Your Own House

Food – Diet – Beverage

A French engineer, André Simonéton, now in his eighties, used the pendulum to select the foods which were crucial in helping him recover from what might have been a fatal illness. During World War I, Simonéton contracted a severe case of tuberculosis. Treatment included five complicated operations after which he was still in a critical condition. Besides the TB, the prescribed diet of rich foods, intended to build him up, had degenerated his liver and caused other side effects. As he lay close to death, he overheard two army medics whispering that he was a goner.

But Simonéton had no intention of fulfilling their dire prediction. He recalled André Bovis' technique of dowsing food for freshness and vitality. As soon as he was able, Simonéton worked out a new eating programme for himself by holding the pendulum over each food and asking if it were specifically healthy for him, a clockwise rotation indicating its healthiness. He discovered that fresh fruit and vegetables, ocean fish and shell fish, olive oil, and whole wheat bread were the most beneficial foods for him. After a few months on this eating regimen, his strength slowly returned, and the TB, liver ailment and various side effects gradually disappeared, leaving him healthy and completely healed. Apparently, his good health has continued. He fathered a child at sixty-six, and another at sixty-eight, and played tennis well into his seventies.

Later, based on his knowledge of electrical engineering and radio, Simonéton was able to measure the intrinsic radiations emitted by different foods and determine which have more nutritive value. He published the findings of his many years of research in 1971 in *Radiations des Ailments, Ondes Humaines, et Santé*.

As a selector of food quality, in the hands of a homemaker, the pendulum could turn an otherwise unbalanced diet into a nutritional delight. The world's population is increasing geometrically, whereas the production of food is decreasing at an alarming rate. In the future, a keen and precisely measured diet may become a matter of survival or death.

Homemakers using the pendulum could, with practise, determine which foods and drinks would be most beneficial for themselves and their families. Doctors, dieticians, nutritionists, and other experts have written volumes of material on well-balanced diets. Unfortunately, these broad suggestions for the masses fail to consider the very particular needs of each individual person. The pendulum can, more or less, solve this problem.

There are basically four methods of selecting the proper family diet. Each has proved to be incredibly successful in actual practice. Having mastered the basic technique of using your pendulum outlined in Chapter Five, you are ready to choose the specific beneficial foods for each member of your family.

The first method involves making an extensive list of foods usually in some kind of order. For example, you may group according to fruits, vegetables, grains, meats, and other categories, or according to alkaline or acid-base foods. You may use whatever classification system that both interests you and meets your personal needs. Many books and brochures provide exhaustive lists of foods under various headings, and will eliminate complications.

The second method is using the yes-no questions: 'Is this food healthy for so and so's diet?' Naming each member of the family successively and then note the pendulum's positive or negative gyration on a card or in a notebook. Usually you will get, possibly to your surprise, quite a varied list for each member, making it difficult when it comes to the actual meal preparation. Most home kitchens are not restaurants where each family member puts in an order. Overcoming this tendency is easy with the pendulum. Simply make a second list, call it the family list, listing the common foods with a positive rotation for every member. If anyone deviates from the family list to any great extent, it may be necessary to cook individually for that person. The half-hour in front of the stove could save hundreds of hours of health care and hundreds of pounds in medical expenses.

The third method involves making pendulum measurements over the actual food products. This can be done right in the supermarket, if you do not mind strange looks from other shoppers and even stranger questions from the manager. Simply hold the pendulum over the fruits, vegetables, cheeses, and other food while asking if it would be healthy for each family member. The pendulum may also be used, as long as you are in the supermarket, to increase your consumer awareness, by asking whether or not the manufacturer's prices are fair. A no answer could indicate an overpriced product.

The fourth method is perhaps more organic and enjoyable. Using your intuition and your knowledge of your family's likes and dislikes, buy the foods and prepare them. After placing them on the table, check the foods out with the pendulum. Either you or each member could operate the pendulum.

Numerous accounts of success using the pendulum for food selection attest to its usefulness. One English lady had an inordinate penchant for fish and chips with lots of fat. Unfortunately, she paid dearly for her pleasure in the form of indigestion. With only ten minutes of pendulum questioning, she discovered that it was the fat not the delicious fish or potatoes which upset her. Now she eats to her satisfaction as long as the fish is dry.

In his book *Dowsing,* W.H. Trinder describes a case from a letter he received: 'A woman in business near here suffered from headaches three times a week, so much so that she thought she would never hold a job. Her lady doctor told her that there was no known cure for that kind of headache. She came to your demonstration and then tried holding a pendulum over her food, with the following results: clockwise, brown bread, boiled eggs, and so on; anticlockwise, white bread, fried eggs, sugar, and so forth. She promptly dieted accordingly. In a day or two her headaches stopped and did not return. The pleased woman told her doctor, who was curious but noncommittal and offered her the pendulum to try. The lady doctor held it over her palm, whereupon it began to whizz around almost as rapidly as yours did in your demonstration. The doctor seemed rather scared and asked 'What is it doing? I'm not doing this." The patient told her that she was evidently a well-developed dowser, and that if she tried holding the pendulum over her food, she might find something of interest. So you can put down to your credit the cure of an unknown woman who had a fixed idea she was going to be forced to give up her job and is now as well as she could wish to be.'

The pendulum can also be used to determine the exact amount of food to be eaten at a meal. Place a very small quantity of food on your plate and ask the pendulum: 'Is this portion just the right size for me?' Steadily increase the amount until the pendulum rotates clockwise. Usually two or three additions are enough to show that the quantity on your plate is sufficient for the time being. Be sure to check the pendulum before taking a second helping. Excessive eating is not healthy. Perhaps a few minutes rest will be enough before second portions, but here again check it on the pendulum. To test drinking water and other beverages like sodas, fruit juices, beer, and wine, simply hold the pendulum over the desired drink and check its quality and quantity with the magic pendulum.

Abbé Mermet did extensive and productive research in the area of testing water. Using the pendulum, he was able to detect water contaminated by organic matter or microbes which was unfit for human consumption. In Berne, Switzerland, during the twenties, the abbé was given the task of finding out why a certain spring water turned yellow with a characteristic taste and smell after each rainfall. The pendulum's movement revealed that a farm two miles away from the spring was the source. A 'perimeter of protection' was placed around the farm putting an end to the spring's contamination. This story is just one of hundreds of successes Mermet enjoyed in locating sources of contamination.

Radiesthetists have done some work in the area of alcoholism. They have been able to determine how much drinking is too much for specific individuals. Also by making a list of alcoholic beverages, placing the name of the person alongside each beverage and asking the pendulum if it is suitable for that person, pendulumists have been able to determine which liquor would most lead to excessive drinking. Much more research needs to be done in this area, however.

The gourmet cook in your life will certainly cook to your tastebuds' delight, using the pendulum as a measuring cup. One hungry researcher found that vegetables stewed together without water – the more watery ones placed below others less watery – give a more positive swing than the same vegetables cooked separately in water. Nutritionists know well that cooking vegetables over a slow fire increases the palatability. It is less destructive to vitamins than boiling.

Radiesthesia may also furnish a valuable key in determining the chemical content of different foods. Think of the tremendous savings in terms of money and work hours in sophisticated laboratory chemical analyses. By simply asking the pendulum, we can determine in a few seconds the amounts of calcium, iron, iodine, phosphorus, and other elements in every food. Hygenists have warned the public in recent years about proper and necessary quantities of these minerals. The pendulum may give knowledge needed without elaborate scientific labels.

Naturally, it goes without saying, that restaurants, coffee-houses, and markets could benefit as much, if not more than, the family unit. It means money in the pocket and happy returning customers to serve only the best of foods. Who knows, the day may come when every chef is taught the use of the pendulum along with the use of the paring knife. Food poisoning will become non-existent and tastebuds will blossom as never before.

The master of radiesthesia, Abbé Mermet, did some fun-loving experiments to detect the alcoholic content of wine. Pendulum, in hand, he would ask the percentage of alcohol. He tells an interesting story of one of his findings. It seems he attended a banquet in a quaint Swiss town one evening and observed, when he was about to sip his wine, that the wine in his glass was slightly lighter in colour than that of his neighbours. Taking his pendulum from his pocket without drawing attention to himself, he found his glass of wine was 9 per cent alcohol while his neighbour's was 11 per cent. He called the waitress over and mentioned the discrepancy, but she flatly refuted the possibility. When the meal was over, Mermet spoke to the manager of the hotel again telling of the discrepancy. But the manager replied, 'All the wine was drawn from the same barrel, and therefore what you say is impossible.'

Mermet insisted, knowing his pendulum readings to be nearly always accurate. The manager agreed to double check and called the waitress over to verify the wine she poured was all the same. The girl blushed ashamedly and confessed that in her laziness to avoid getting another bottle from the cellar, she had filled the last bottle with some water, which was the jug from which Mermet's wine was poured.

Missing and Hidden Objects

All of us have experienced the quiet morning that turns into a mad hunt for something we have lost, the car keys, a sock, or whatever. Usually our first reaction is overly emotional: either we did not put it where we thought we did or we blame someone else for its present invisibility. Our next reaction usually has us hurrying and scurrying all over the house or flat.

Next time this happens, stop! Take out your pendulum and find the object. Point to each room of the house and ask, 'Is the missing object in this room?' Continue till you locate the room or hallway, or place. Then you can check each section of the indicated place until the pendulum reacts with a positive swing. Your car keys, or whatever, should be there.

Another method of locating a misplaced item involves picturing the object clearly in your mind and asking the question, 'Is the lost object in this direction?' Continue to move in a 360 degree arc as you ask the pendulum. When it swings positive repeat the same process in another part of the room. The point where the two positive directions cross is the location of the missing object.

There is a remarkable story about the former president of the American Society of Dowsers, John Shelley, and his ability to locate a hidden object. During the final week of training at the Pensacola, Florida naval air station, he amazed his fellow naval reservists by locating his pay check. It seems his buddies along with the paymaster hid the check inside a giant two-storey naval building with a few dozen rooms branching out of the long hallways. Using the pendulum, Shelley found it quickly and without difficulty.

Handyman's Pendulum

Naturally, using only the pendulum to discover the cause of a household problem is not enough. You may locate an electrical or plumbing difficulty, but you still need the skill to correct it. We strongly suggest, therefore, that you do not repair devices about which you know nothing. Leave it to the experts or to someone knowledgeable about such things.

However, the pendulum can help you solve minor household problems. When a lightbulb needs replacing, you can find out the exact number of watts the lamp should and can hold. When trying to find a two- by-four behind a plaster wall on which to hang a picture or photograph, you can use a pendulum. To find out when to change the vacuum cleaner bag without opening the machine, check with the pendulum. There are a thousand and one handyman repairs around the house which using a well-operated pendulum can make easier. It should be in every home tool-box along with the hammer and screw-driver.

One Arizona pendulumist has successfully used the pendulum in car repairs. He has used the magic pendule to help him adjust the points, select a bad plug without having to remove them all, and to adjust the carburettor as accurately as the best diagnostic machines. In this day of outlandish car repair prices, home car tune-ups – using a pendulum – will save you money and give you knowledge which many mechanics take years to learn.

Houseplants

In recent years houseplants have become the rage. City dwellers especially have started collecting indoor plants. Along with the increase in houseplants have come numberless books explaining how to water them, talk to them, feed them, re-pot them, and so on. Again the pendulum can cut down on the time and money. Each time you water the plants, check with the pendulum to determine how much

Figure 12. Pendulum and plant communication.

water you should give each particular plant.

Also you can ask if the sunlight is too much or too little for a plant. You can find out when a plant should be re-potted, what kind of soil it would enjoy, and the best size of the pot. You can easily find out whether or not a plant is happy and healthy by holding a pendulum over it, as in figure 12. If the gyration is positive, the plant is happy, if negative, it may be, and probably is, unhappy. Next, you can discover the root of the unhappiness. Perhaps, the sound pollution, air pollution, or water pollution is bothering it. Maybe the emotional state of a family member makes it droop. Possibly the plant's owner does not harmonize well with a particular plant and thus it should be given a more harmonious owner.

A friend of ours, who keeps approximately one to two dozen houseplants, finds that by using the pendulum she can determine the exact spot in the house each plant likes. First, she uses her intuition and places the plant where she feels it would be best. Then she consults the pendulum while moving the plant until a clockwise swing reveals the best spot. Also she has made a surprising discovery – certain plants do not get along with other plants! Being next to each other retards their mutual growth and happiness. Be sure to check each plant's relations. Who knows maybe your favourite plant has an unhappy life, drooping from lack of love.

Home Decoration

Modern day depth psychology has rediscovered what ancient sages knew – that the environment has a powerful influence on our psyches. The pendulum can unlock the mysteries of home decoration by telling us what colour schemes should be used in each room to best enhance the family life. The wrong colour scheme in the bedroom may ruin the love life. Inappropriate furniture in the living room could make you uncomfortable. Selecting your carpet, paints, furniture, and other home fixtures, with a pendulum, may make the difference between a happy family or a broken home.

Home Location

A happy home may depend as much on locality as on interior decoration. Many radiesthetists are able to determine with amazing accuracy what country, county or town would be the best for a family, couple, or individual person. The technique is easy and takes only a few seconds. Write on a slip of paper the name of the locality in question, concentrating and focusing your mental energy on it. This process works much like tuning a radio to a certain station. In this instance the mind is focusing on the place and centring the vibrations of the place on the name. Next, ask, 'How well would so and so's vibrations harmonize with such and such a place from the viewpoint of happy living, good health, and monetary success?' Watch the strength of the rotation which will indicate the degree of harmony or disharmony.

In recent years, radiesthetists have become aware of harmful earth rays at certain places. Building a house or other edifice at such a spot could be disastrous to the health of those living there. Here is one man's chilling account:

> Not long ago, I went to live in a house known to be on clay soil. I had not been there long before I began to lose my energy and became very run down. In fact, I found it quite an effort to move

around. Later on, I had a somewhat remarkable experience. I woke up early one morning with my hair standing on end, and I had a cold, clammy, tingling sensation down my spine. I felt limp, miserable and exhausted. Moreover, all the tone seemed to have gone out of my nerves and muscles, and what was worse, the symptoms persisted from that day.

It so happened that an experienced lay dowser was visiting me at that time, and I took her up to my bedroom. She told me that from a radiesthetic point of view, she did not like the house at all, and that the worst radiations were close to my bed. This was confirmed by another radiesthetist, who obtained similar findings by the map dowsing method from a rough sketch of the house I sent him. He told me that they were some of the worst radiations he had ever come across.

No doubt in the near future, architectural and engineering firms will consult pendulumists to detect these harmful rays.

Your Children

Mothers and fathers will find that a little expertise on the pendulum will go a long way toward helping their children. Measurements may be made to determine their potentials and limitations. Both parents and teachers can be aided in better understanding the child. If there are any retardation or learning disabilities present, the parents and teachers can select the best methods of schooling, therapy, and overall handling of the child's problem.

The teachers responsible for a child's education should be carefully checked. The pendulum can measure on a scale from 0 to 100 the quality of the teacher's ability relative to your child. By finding out the degree of harmony or disharmony of your child with a teacher before the year begins, you can prevent many months of possible unhappiness.

In family relationships, the pendulum may do much to bridge the generation gap. By checking the pendulum on important issues in the presence of the child, a parent may avoid becoming 'a bad guy' if it rotates a no answer. The child is more likely to accept this impersonal reply. There is no one to blame. You will discover that there are innumerable possibilities for using the pendulum in raising children, from determining the bedtime to how they are doing in school to friction among siblings, indeed, in every imaginable situation.

13

Superpendulum

After spending a great many hours of consistent practice in using your pendulum you will eventually reach the point at which you will be ready to explore its more advanced uses. In the next few pages we will present some of the pendulum's more advanced pendulum uses. Most are out of range of the newly initiated but actually are not as difficult as they may appear to be at first sight. A little effort at repeated intervals will bring you much insight and greater pendulum skill.

Map Dowsing

Map dowsing has become widespread in radiesthesia circles and societies. Even the United States Marine Corp instructs their mine experts in the technique of detecting the enemy's mine fields, using a map and a pendulum.

The technique is simple and in most cases successful. First, obtain a map of the district and place it on a table or any other flat surface. Second, hold the pendulum over each of the four sides of the map representing north, south, east, and west, and then in the centre (see Figure 13). Over each area, the question: 'Is what I'm looking for in this direction or area?' When the pendulum responds by rotating clockwise, it indicates that the buried treasure, water, or mine field is in that general area. The next step involves a narrowing down process. You must ask the pendulum specific questions; for example, 'Is it located in such and such a town, such and such a street, such and such a corner?' Eventually, you will be able to locate the desired substance or person as accurately as radar locates aircraft.

There are endless accounts of map dowsing in radiesthesia texts. Some are documented and therefore carry more weight than a mere claim of success. One such account describes the map dowsing talents of an Englishman, who wrote of his findings in the *Journal of the British Society of Dowsers*.

The Englishman, Mr Clark, asked a police sergeant, whom he had just met, to write down exact directions to reach a desired destination

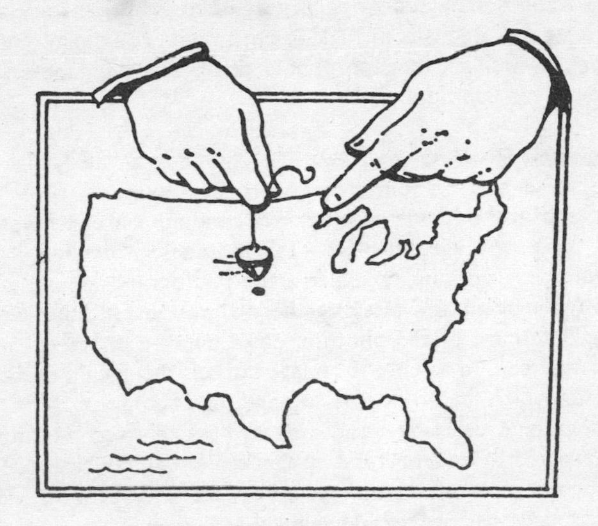

Figure 13. Map Dowsing.

by car. Without reading the directions, Mr Clark drove off. A short while later, he returned and described the exact route, explaining how he had used a map and his pendulum to find the directions.

A French map dowser, Joseph Treyve, received a packet containing two stones, picked up somewhere in France. The sender requested that he determine the exact location where the stones were found. Within a few minutes, he wrote down his answer and mailed it: 'These stones were picked up three kilometers from Eysies in Dordogne by a young, fair, and pretty woman.' He was correct in every detail.

Hidden treasure buried and lost, ships with chests of gold coin lost at sea – these fortunes may now be found by map dowsing. Abbé Mermet had some experience as a treasure finder. A letter in 1934, he received with congratulations on his find.

> About two years ago, I consulted you about making a search for gold on my property. You indicated a place where there was 400 francs in gold. And the place you had marked on the map which I had sent you was exactly where my safe had been put containing twenty-three gold coins. (signed) Georges Otto.

Prospectors, geologists, and mining engineers, may some day carry pendulums together with their maps. With gas, petroleum, and other natural resources dwindling each year, there will be a dire need for discovering new deposits of fuel. The pendulum could pinpoint precise locations as well as the appropriate depth at which these valuable substances can be found.

Photographic Dowsing

The Chinese say, 'A picture is worth a thousand words', but a photograph in the hands of an expert pendulumist can reveal much more – a person's age, birthday, basic attitudes, profession, and other information. According to experts, the photograph acts as a tuning device by the occult law of sympathetic vibrations. In other words, by holding the image of the photo in one's mind, a direct link is made with that person. In a sense the image attracts the distant radiations of the person which then condenses around the photograph.

Photographic dowsing would reveal anyone's secret attitudes or intentions. World leaders could be checked on a daily basis to find out their mental-emotional stability. Even the members of the U.S. Congress could be checked individually and collectively.

During World War II, an Italian radiesthetist held the pendulum over the photographs of his two sons to determine their whereabouts and physical condition. He knew before receiving any official word that his youngest son had drowned at sea.

In contests, professional as well as amateur photographers could submit only the best photos by selecting them with a pendulum. Newspaper editors looking to sell the greatest number of papers could choose the front page shots using a pendulum. Also film editors who spend long hours splicing together a motion picture, would be able to pinpoint the exact frame to splice.

Photographic dowsing as yet has had little, if any, constructive research, or at least none that we know of. Pendulum users may want to experiment in this area and share their results. We have found in our experience in photographic dowsing that tuning in to the person while holding the pendulum over a photo gives a more accurate character reading of that person than just holding the image in your mind.

Pendulum Charts

Without a doubt, pendulum charting is one of the most useful radiesthetic applications to be opened up in recent years. Writing the words 'Yes/No' on a piece of paper to use as a diagram can be helpful in determining an answer. Somehow having a visible chart to work with aids the mental concentration, which definitely becomes sharper, clearer, more precise, and increases the success of the pendulumist.

The Yes/No chart can be used in practically any situation to determine health or ill health, the genuineness or phoniness of a gem or painting, or its value or worthlessness, or any other quality. The chart is extremely easy to use. Simply hold the pendulum over either the Yes or No and ask your question. If the pendulum rotates clockwise over Yes, your answer is obviously, an affirmative one. Now, to double check a Yes answer, hold the pendulum over the No. If it rotates counterclockwise, indicating a No, then your affirmative has been substantiated.

An Indian pendulumist, Ben Bhattacharyya, devoted a sizable portion of his book *Magnet Dowsing* to the use of pendulum charts. In fact, some occult book shops sell a deck of pendulum chart cards which cover such areas as colour radiation, planet radiation, health index, and brain radiation.

Also a number of charts have been used in medical radiesthesia to discover impending ailments. An Italian, Pietro Zampa, has successfully used a diagram chart of the human hand, having the numbers one through nineteen written on it in specific areas. A negative gyration over any number indicates some difficulty. The value of this method, when used professionally, is that it can detect radiations of diseases still in their initial stages without elaborate testing and exorbitant fees.

The Pendulum Swing

So far we have mentioned the two basic swings of the pendulum, clockwise for 'Yes' and counterclockwise for 'No'. However, for any advanced work with a pendulum the Yes/No rotations are not adequate. The advanced pendulum user will want to know the type of the positive or negative rotation; specifically, the meaning of the radius of the swing, the intensity of the swing, and the duration of the swing.

For example, if we ask the question, 'Are Mr and Mrs John Smith compatible in their relationship as husband and wife at present?' If the pendulum rotates with a positive swing but in a small radius, say an inch or so, and the intensity or speed of the turns is very slow, this type of swing indicates compatibility but only in a very limited way. The romance, vitality, and glamour of the relationship has probably worn thin and needs to be renewed or strengthened.

Now, say things have changed between our imaginary couple. Mr and Mrs Smith have been fighting, squawking and bickering like hens in a roost. When we check their compatibility using the pendulum, it not only rotates with a negative swing but also revolves in wild gyrations that have a radius of six or seven inches and a speed sufficient to send the pendulum flying out of the pendulumist's hand, maybe divorce is just around the corner.

In our years of pendulum use, we have come across other sophisticated swings. Every so often the pendulum may rotate in both directions; that is, it may start rotating clockwise and then stop and begin rotating counterclockwise. Depending on the question, it may mean any number of things. Assume that in a compatibility check with Mr and Mrs Smith their relationship alternates consistently through positive ups and negative downs. In matters of business investments it may show a considerable profit but unknown obstacles which put a hole in the stomach while filling the pockets.

Finally, a non-swing response of the pendulum after asking a question indicates that the question was poorly worded, a foolish question, or a question that should not be asked at all. If it is a poorly worded question, keep re-wording till the pendulum gives the go ahead. 'How many camels can pass through the eye of the needle?' or 'Should you start eating an ice cream cone from the left side?' are foolish questions that you should drop altogether. Spend your time asking more productive questions and the pendulum will work for you.

Also drop questions that should not be asked in the first place. It seems there are some answers which are not meant to be known till an appropriate time. For instance, knowing when and how someone will die may be none of your business and if your reading is accurate by subconscious telepathy, you could harm the person being checked.

Time and the Pendulum

What time of day is it? If there is no watch or clock available and you are carrying your pendulum, you can find out. Simply ask the pendulum while allowing it to rotate counterclockwise 'Is it one, two, three ... o'clock?' The pendulum will start rotating clockwise when you get to the correct time of day.

A second method of telling time requires a water glass or any object that has a straight side. Hold the pendulum about two inches from the side and ask the time. The pendulum will suddenly swing striking the glass or the vertical edge of the object. Count the number of direct strikes, and you will be amazed to find that the number of strikes gives the nearest number on your dial.

The pendulum can do what a regular clock cannot; that is, it can determine how long it may be from the present hour until something you expect to happen actually does occur. Say, for example, you are expecting an important business call on a big money deal, or a call from a good looking guy you met the day before. You know approximately when the call will be made, but you want to know the exact time so you are not caught off guard, sleeping, or taking a shower. Using the same method with the glass or vertical object, ask how many hours it will be before the call comes. If the pendulum does not respond, it may be because the call will be made in a matter of minutes. In this instance ask how many minutes before the call comes, counting the strikes as minutes rather than hours.

This method of telling future time can be expanded to measure days, weeks, months, or years. One couple eager to find a new home wanted to know how long it would take to find a desirable place. The pendulum indicated seven weeks. It was seven weeks to the day when they found a suitable small house in a wooded area.

According to occult theories, the subconscious mind does not measure time in the same way as the clock. It is far more plastic, extending far into the future and into the past without setting up limitations such as our usual time-space relationships.

Much research needs to be done in estimating time. We suggest that you keep careful notes of your questions and answers. Not only is this good scientific procedure but also you may forget in the course of daily living a question you asked months or even years earlier.

Besides estimating future events you can recall or find out exactly when events occurred in the past. For beginning exercises choose a person you do not know, a casual aquaintance, a friend of a friend, or perhaps someone at work, someone whose age you do not know. Hold the pendulum next to the glass as described in the time exercises, or if you prefer, write down the years and check each year until you get a clockwise swing. Always be sure to check your answer until you achieve a certain measure of consistent success.

This method of pinpointing past events will be invaluable to you personally. Everyone has gone through trying situations, traumatic experiences, and some kind of tragedy. The pendulum can help you to digest these past experiences.

We know that psychologists and psychiatrists survey your past, digging for attitudes brought on by certain experiences. Our childhood difficulties can determine to a very considerable extent our present circumstances and conflicts. These past experiences need digesting, dissolving, and discharging to liberate the energy so necessary for the difficulties and pressures of modern living.

Usually the psychiatrist and psychologist spend endless hours costing you plenty to uncover only a few of the pressure points. The pendulum can give you this information in a matter of minutes and without cost. Holding the pendulum over each word in the chart, ask: 'Is this an area which needs digesting?' Make a list of the pressure points and then at a time when you can sit down in relaxed way, jot down on a piece of paper those memories you recall in that area. Do not analyze and do not let your emotions get carried away. Do it for just a few minutes maybe ten or fifteen to begin with. Check the pendulum for the exact amount of time to spend. When finished rip up the paper and throw it away. This discharging technique will help you to release the locked up energies and open you to a greater understanding of the past conflict so that in present you handle a similar situation more easily.

Radionics

A separate field of radiesthesia, radionics, operators use machines to pick up radiations instead of a pendulum. The operator sits in front of the machine which has one, two, or three horizontal rows of dials each with number settings. Usually one corner of the machine has a rubber surface on which the middle finger is placed. When the dials are properly set, in response to a question on a specific situation, the feels a tuned reaction when the finger drags on the rubber. This phenomenon is known as a stick.

One of the first radionic machines was constructed by Dr Albert Abrams whose work as a pioneer in radiesthesia we mentioned previously. His machine was called the E.R.A., Electronic Radiation of Abrams. Most of his work took place in California between 1900 and 1923. In 1924, he ran into strong opposition to his work by two powerful groups, *The Scientific American* magazine and the American Medical Association. They called his machine 'the magic black box of Abrams', and said it should be removed from the market.

Many legal battles and ostracisms have occurred over radionics since that time. In the 1960s one of Dr Abrams co-workers, Dr Ruth Drown along with her staff were thrown into jail by the Pure Food and Drug Administration. Still, research is going on in many countries. A number of books and magazines have covered the topic thoroughly and are listed in the Bibliography.

Note, however, that in the September-October issue of *The Psychic Observer*, Ms Frances Farrelly, who has been in radionics work since 1940 said, 'You can use the box (machine) as a tool if you want, but it is not doing a blessed thing.' In other words the operator receives the vibrations not the machine. As she goes on to say, 'It is far better to learn without instruments, without the expense this incurs (from $500 to $5000). It is far better to train your own mind because the mind is the diviner.'

Taking Ms Farrelly's statement to heart, we recommend the pendulum over the radionics machines. The pendulum is simple, easily accessible, and with diligent practise easy to use. Also it teaches you to focus your mind out of necessity, which will help you develop your potential to a greater degree than if you allow a machine to do the work for you. Another benefit that comes with learning pendulum technique is acquiring the ability to be emotionally calm and to have a neutral attitude when asking the pendulum a question. Emotional objectivity can benefit you in every area of your life, in all relationships at work and at home.

14

The Pendulum by Day and by Night

At the recent fifteenth convention of the American Society of Dowsers noted French radiesthetist Abbé Jean Jurion described the present day pendulumists of France. Believe it or not, in his country there are enough full-time professional radiesthetists to have organized the National Union of Radiesthetists.

Its members have united as a group of professionals to defend the individual members against attack and to inform the general public about the wonders of radiesthesia. The union also publishes a magazine, *Radiesthesie Magazine*, detailing the extensive activities, the considered opinions, and dedicated research of its more than one thousand members.

France is not the only country to accept the pendulum's incredible possibilities. Today, in the Soviet Union, pendulum divining goes under the name 'The Biophysical Effects Method', or 'BPE' for short. Their main focus is on the discovery of national resources. But they are also exploring the pendulum's uses in discovering the unknowns of man: why there are certain diseases, how they are caused, and how they can be prevented.

In the United States, radiesthesia has not been accepted by the scientific community or the general public. Still, thousands of individuals as well as small groups are experimenting and practising the art of pendulum divining.

My co-author and I each carry a home-made handcrafted pendulum at all times. We never know when we might be called on to do a reading. On numerous occasions I have used the pendulum to help friends at work in a large New York department store. They have been amazed as well as awakened to the use of the pendulum for measuring unseen energies, vibrations, and radiations.

More often than any other thing, they ask me to check the relationships or compatibility between two people. As we described in the chapter 'Romantic Radiesthesia', when two people meet, their auras, or more scientifically speaking, their force fields blend (or clash), interact, and inter-react, setting up a condition that can be tuned in to by using a pendulum. So far I have had uncanny results in this area. Just by knowing only the names of the two people and no other information about them, my pendulum reading of their compatibility has been 90 to 100 per cent accurate.

Recently, I befriended an airline stewardess and explained to her how the pendulum works. One night she called me to discuss a difficulty with her boyfriend. She must have asked at least fifty questions including: 'Is he home now? Does he love me? Can our love last? Does he love any other women?' A couple of days later she learned the answers when her boyfriend returned from a visit home and let her know the depth of his love for her and his disinterest in any other woman. Approximately two weeks before this over-the-phone pendulum reading, the stewardess had consulted two gypsies, one from New York and another from Boston. The pendulum reading agreed with everything the two psychic gypsies had to say about the boyfriend.

Of course, not everyone greets us with acceptance when we show them the pendulum. Most pendulums run into the sceptic, the scoffer, and the condemner. We have had numerous such encounters recently. Here is a description of one typical incident.

The gibberish hum of two hundred or more people talking with their mouths full or partially so was the dominant sound in the lunch room. I paid the cashier for a tunafish salad sandwich, a pickle, and coffee.

I looked across the lunch room, my eyes scanning for the familiar face of a friend. Seated by the windows on the far side of the room I saw Tony munching on a sandwich. We had worked together in the comforter and blanket department during the turmoil of the Christmas season. I sat down opposite him.

'Hi, Tony, what's happening?'

'Same old thing. The old bags are squawking us usual. And, of course, they're still asking me where the ladies room is. How's your book going?'

'Writing away, should be finished in another month or so.'

Just then, one of Tony's friends came over and sat down. Tony introduced us. Bill was his name. Tony told him I was writing a book on pendulums.

'You're writing a book on pendulums, like a pendulum in a clock? What can you write on that?' asked Bill in a harsh tone.

'No, I'm writing about a device which picks up, or tunes into, vibrations.'

'What, are you crazy? A device that picks up vibrations. Ha, ha, ha

– right. What do they look like?'

I knew I had a real doubter on my hands but I went ahead and pulled the pendulum from my pocket. Bill looked at it and his eyes revealed his inner feeling of ridicule.

'Show us how it works,' he demanded in a challenging tone.

I proceeded to draw a cross on a piece of paper, showed Bill how to hold the pendulum, and told him to look at the pendulum and concentrate on moving it along the vertical line, with the power of his mental attention.

He laughed and began. I immediately saw that he was resisting the possibility and that he was thinking it could not work. His hands and body became rigid. He did not even really try concentrating on it. Instead, after less than a minute, he said, 'It doesn't work. My mind can't move it. Besides if it did move, how do you know my fingers aren't doing it?'

You have to understand the kind of negative emotional force these scoffers and condemners emanate from their tone of voice and feelings. They try to make you feel stupid, inept, crazy, insane, and ridiculous. As if they are on the side of science and 'know all'.

Some words of advice to pendulumists and radiesthetists when confronted with this situation. Do not try to prove the pendulum works to the empirically minded types. You cannot prove anything to the unreceptive mind. Two apples and two apples are not four apples to one who refuses to see. If you try to show them how it works, nine times out of ten you will fail since their negative attitude affects the pendulum's reading.

If it so happens that they begin to attack you for believing the pendulum works, do not take it personally. If you react and try to defend your skill with some verbal argument, he will think he has you. It is much wiser not to identify personally. For example, I have learned, for the most part, to divert the attack by saying, 'Experts have reported that the pendulum works although I, myself, cannot prove it one way or another.' Usually, this disarms the sceptics and their tension begins to relax.

My co-author and I have experimented with some of our pendulumist friends in a relatively new area of pendulum work, doing pendulum readings over the telephone. A week does not go by in which we are not consulted over the phone to give an answer to a pressing question on romance, family problems, or business deals. It is perhaps more advisable to have the person physically present since the psychic connections of thought and feeling are usually stronger and thus the reading more accurate. Yet the laws of the inner planes bind us together through subtle invisible threads and a pendulumist who has enough experience and training can obtain excellent accuracy over the phone.

Telephone reading, we must point out, demands specific steps to ensure accuracy. Most importantly, the person asking the question should be in a proper sitting position, upright with feet flat on the floor. The pendulumist should also maintain this position. If either one lies down or sits with his legs crossed, it will break the psychic or inner circuits automatically, and reduce the quality of the reading.

Those of you who decide to take up the pendulum and use it as a tool in guiding your way through daily difficulties should adopt an attitude that is balanced. Specifically, you should avoid the extreme approach that the pendulum is the *only* answer or the *only* way to steer a life course. A more balanced approach would be to see it as a part of the whole picture, accepting other methods and disciplines equally valid under other conditions. This more balanced approach will ensure a greater accuracy in your readings and leave you open to new possibilities, not only with the pendulum but also in other disciplines so that you can improve your level of being – the quality of your thoughts, feelings, and actions as well as the inner connections to your deeper self. Most likely, you will find this approach bringing you into contact with a larger number of people who are willing to accept not only the readings you give but also you as a person.

In addition, this balanced attitude will help to remove the taint of negative occultism which many people automatically have when you show them the pendulum for the first time. Beginners as well as adept pendulum users must remember that to others not familiar with the pendulum, it has a strange and even a ridiculous look. Educating others involves understanding, patience, and a willingness to gracefully cast off abuse and criticism. Let your work speak for itself. If you are accurate a great percentage of the time, gradually others will accept your skill and consult you more often and even learn to use the pendulum themselves.

Finally, we would like to remind anyone interested in becoming a qualified expert pendulumist that it requires a period of apprenticeship and constant practice to develop the reflexes in nerve, muscle, and mind necessary to be skilled. Just as with any other acquired skill such as writing or typing, it takes patience and training to enable the skill to grow into an innate and acquired ability. Keeping this in mind, almost anyone can develop pendulum power to one degree or another; and we must add, it is a power which must be under control, well-intentioned, and precisely placed.

15

Visions of the Future

One cannot help wondering what the implications are for humanity as a whole as more and more people become proficient and expert in the use of the pendulum. The development of the radiesthetic faculty, or sense, far transcends the importance of any mere technological breakthrough. For here we have a breakthrough in the evolution of man. We have the full development of a new sense, the limits of which seem infinite. We have all seen what an advance such as the discovery of the steam engine brought to civilization. It gave birth to the Industrial Revolution and completely transformed society. But the steam engine and other technological advances were merely external. They were effects of an inner change in some members of mankind who had trained and developed their intellects. The scientific method came first, technological advance was a result.

To get some idea of the scope of what we are talking about, imagine yourself living in a society and civilization built around a race of beings possessing only four senses. They have all our present senses but lack sight. Try to visualize such a society.

People would live in a world of eternal night. There would be no light, no colour, no literature as we understand it. It is doubtful whether any kind of science or technology could develop with no sight. Perhaps the obstacles to building precision tools and machines would eventually be overcome, but it would take millions of years longer. It would be a primitive and dark society.

Now imagine that through evolution some members began to develop the sense of sight. And these members saw how this sense could be developed in others. Would you agree that a sightless society would be utterly transformed? The beings with the sense of sight would be like a new race, a new step in evolution, like the caterpillar becoming a butterfly, they would be a different order of being. And as more and more members of that society became sighted, the old institutions, moralities, modes, and habits of thought and living would utterly disappear.

Hitherto, insoluble problems that taxed the greatest minds of the old civilization would be as nothing even to infants of the new breed. With the gift of sight most of the problems would just disappear. They would be meaningless and irrelevant.

A similar situation confronts us today. Until now the great majority of humans have functioned with only five senses. Our perception of reality has been limited, and because of this, we have built a civilization that reflects limitation. In this, the latter part of the twentieth century, we are confronted with so many seemingly insoluble problems that most of humanity, including our intellectuals and leaders, are in a state of confusion, fear, and despair. As S.I. Hayakawa says, 'When the intellectuals of a society are confused, the general populace follows suit.' For the intellectuals are to a society what the brain cells are to a body.

Now try to envision our culture when the science of radiesthesia and radiational physics becomes accepted, and a good portion of the people are skilled in their use. Most crime would just disappear. Who is going to rape, rob, or kill when detecting their identity and apprehending them is almost a certainty by detectives skilled in radiesthetic and teleradiesthetic techniques. Holding the pendulum over a map and a fingerprint, or an article of clothing, or a police artists' photograph, an expert operator can track down the criminal in any part of the globe. Who is going to write phony checks when any trained person can just hold a pendulum over it and spot it immediately?

We are faced with a tremendous justice problem because of case overloads on judges, cumbrous and slow courtroom procedures, and the right of anyone to demand a trial by jury. Get enough judges skilled in the use of the pendulum, and they will be able to determine instantly whether or not the man is guilty, or whether or not a witness is lying. Eventually, the pendulum can be used to mete out sentences, not by precedent or law books written many years ago, but by the dynamic and specific structural necessities of the present moment. If the criminal is capable of rehabilitation, the pendulum will pick it up immediately, and certain measures can be taken to send this person to a rehabilitation centre. If the pendulum reading indicates that the criminal is not capable of being rehabilitated (these are exact vibrational measurements), then he or she could be sent to a different kind of institution, designed to keep him or her out of society's way.

Because of our lack of perception, we treat all criminals more or less equally. The idea being that everyone has the right to a second chance. And the result is that many hardened criminals return to the streets and repeat their crimes. From an energy point of view, this is wrong. Everyone is not always entitled to a second chance, especially if that someone is a permanently destructive force to society. This can be measured very accurately, and verified empirically.

Ancient Egypt, the civilization that produced the pyramids and which was the font through which the Western world received most of its true science and philosophy came the closest to producing a perfect society for its time.

There, justice was administered by priests of the temples, trained seers, and diviners, who underwent years of rigorous training in character and skill. In other words, they were perceptive beings. Criminal actions and disputes were brought before them which were handled efficiently and with cool measurement of the solution that would be best for society and ultimately for the criminal's own higher growth. These ancient mediators of justice knew that just as a scorpion or shark must be true to its own nature which is predatory, so there are certain human beings with states of consciousness that can only – must only – manifest destructively.

Two of the greatest problems that consumers face today are consumer fraud and false advertising. These also could become crimes of the past. When people get their extrasense or supersense (to use a word coined by Christopher Hills) trained, they will be able to immediately check to see if a product is good or not. When advertisers realize that the public can no longer be fooled, they will stick to competing on the basis of quality rather than brainwashing. Food manufacturers will think twice before they saturate their products with harmful additives. When they know that the populace can easily check the quality of the products, they would be foolish to continue this practice.

Domestic harmony will increase because inharmonious people will know that they are inharmonious and will stay away from long-term committments. People will be happier in their work because they will have the means to select vocations in line with their inner structures and abilities.

With more widespread use of the radiesthetic sense, employees will be selected on the basis of both skill AND compatibility, and this will take out much of the frustration and nerve jangling that causes so many people to hate their work. Productivity will increase and so will quality, and so will inner peace and poise.

People will learn the importance of finding their own rhythm. With the pendulum you can check how long to work before taking a break, which is a variable that depends on each person's particular needs.

With more domestic tranquility, career satisfaction, rhythm, better dietary habits, better and more harmonious relationships, and less fear of crime, most of the causes of illness and disease will disappear. The health of the nation will soar. Disease will be seen for what it is, a violation of one or more of the energy laws of life.

Radiesthesia and radiational physics will completely revolutionize medicine. Doctors will be able to diagnose in minutes, and prescribe treatment even for the most serious diseases. They will have more time to get involved with prevention and with teaching people how to live in harmony with the natural-order process we call God. For treating the cause and the total human being is the true function of the healer. At present, physicians are so overworked and overloaded that this aspect of their craft is functionally impossible, and they usually just deal with alleviating the pain or symptom rather than discovering the cause.

Society's customs of doing business will change. The marketplace will change from a jungle to a cultivated garden. Very little cheating will take place. How can a supplier degrade his merchandise when the customer can check the quality of it in seconds? How can a salesman lie to a customer when his ability to deliver on those promises can be checked almost instantly?

The arts and the sciences will enter a new renaissance. People will see the arts for what they are: forms of beauty to transmit power. The artist will be looked on with the same respect as the engineer or physicist. Through his vision and craft, the artist is the creator of the vehicles through which the universal energy can show the way to the future. The artist and the scientist will be to society what headlights are to a car. They will be beacons lighting up the road ahead so that those responsible for directing and steering the nation will be able to see where they are going and make the proper adjustments.

All that is vicious and vile in art, all that degrades and distorts the minds of the masses will disappear. For works of art, too, can be checked radiesthetically. Art and science feed the mind and emotions, just as food feeds the physical body. To be healthy, a person must be as careful of the quality, and digestibility of his mental-emotional foods as he is about his physical foods. For poisonous ideas (ideas that are false-to-fact) and negative emotions poison the inner bodies and create psychic and physical disease. Unhealthy people cannot create a healthy society.

With the tool of radiesthesia, science will be able to probe hitherto unprobable dimensions. Scientific methods will be brought to spiritual realities, and science will enter a new period of growth and development. New sources of power will be discovered which are safe and which do not rape the environment.

It is as difficult to envision what will happen to science given this new instrument of measurement as it was for Galileo to envision the implications of the first telescope. Since science is based on the observation and ordering of empirical data, a new sense, which is another opening or window to the environment, must increase our ability to observe, measure, and accumulate data in geometric proportions. It must revolutionize science.

In the science of radiational physics, science, religion, and art meet. The inner or esoteric meanings of the ancient religion become evident. Humanity will see religious ritual for what its true function is – the imposition of a higher and truer pattern on the mundane things of life. Society will view science as a process of knowing, art as a process of applying this knowing to the creation of beautiful and harmonious forms, and religion as combining both science and art in handling our everyday relationships and lives. Religion, in our new society, will become the science-art of living according to our natural patterns.

With the growth of the consciousness of the race must come a change in the ethics of the race. No longer will people be enslaved and tyrannized by guilts and fears stemming from outmoded ethics formulated thousands of years ago, distorted and twisted with the passage of time from their original meanings, applied out of context to a different space-time continuum, and interpreted by leaders whose perception of reality was based on only five senses and thus faulty and incomplete. New codes of morality and new systems of ethics will be formulated based on the laws of energy as perceived by a people who now have the ability to see and understand these laws.

All of these, of course, will not happen overnight, nor perhaps in our lifetime. But we can begin now to re-orient our consciousnesses to the dynamics of the energy-process universe, which is God in manifestation. As Isidore Friedman says, 'Do not look for any other god.'

The pendulum is a tool – one tool, perhaps, out of many – which can lead us to the portals of perception.

Bibliography

The American Dowser (60 issues). Vermont: American Society of Dowsers, 1960-1975.

Anderson, Mary *Divination: How to Use Unusual Methods.* Aquarian Press.

Abrams, Albert. *New Concepts in Diagnosis and Treatment.* San Francisco, CA: Physico-Clinical, 1924.

Archdale, F.A. *Elementary Radiesthesia.* London: British Society of Dowsers, 1966.

Askew, Stella. *How to Use a Pendulum.* Mokelumne Hill, CA: Health Research, 1955.

Baerlein, Elizabeth and Dower, Lavender. *Healing with Radionics.* Thorsons, 1980.

Barrett, Sir William, and Besterman, Theodore. *The Divining Rod.* New York: University Books, 1968.

Baum, Joseph. *The Beginners Handbook of Dowsing.* New York: Crown Publishers, 1974.

Beasse, Pierre. *A New and Rational Treatise of Dowsing.* Mokelumne Hill, CA: Health Research, 1975.

Bhattacharyya, Benoytosh. *Magnet Dowsing.* Calcutta, India: Firma K.L. Mukhopadhyay, 1967.

Bhattacharyya, Benoytosh. *The Science of Cosmic Ray Therapy or Tele-therapy.* Calcutta, India: Firma K.L. Mukhopadhyay, 1972.

Mary Everest Boole, Collected Works. Edited by E.M. Cobham. London: C.W. Daniel, 1931.

Burr, Harold Saxton. *Blueprint for Immortality: The Electric Patterns of Life.* London: Neville Spearman, 1972.

Cameron, Verne L. *Aqua Video – Locating Underground Water.* Elsinore, CA: El Cariso Publications, 1970.

Cameron, Verne L. *Map Dowsing.* Elsinore, CA: El Cariso Publications, 1971.

Cameron, Verne L. *Oil Locating.* Elsinore, CA: El Cariso Publications, 1971.

Clement, Mark. *The Waves That Heal.* Mokelumne Hill, CA: Health Research, 1963.

Coblentz, W.W. *Man's Place in a Superphysical World.* New York: Sabian Publishing Society, 1954.

Cooper-Hunt, Mjr. C.L. *Radiesthetic Analysis.* Mokelumne Hill, CA: Health Research, 1969.

Copen, Bruce. *The Practical Pendulum.* Sussex, England: Academic.

Copen, Bruce. *Radionics.* Sussex, England: Academic Publications, 1974; Sussex, England, 1974.

Copen, Bruce. *What Radiesthesia Is!* Highfield, England, 1953.

Davis, Albert Roy, and Rawls, Jr., Walter C. *The Magnetic Effect.* Hicksville, New York: Exposition Press, 1975.

Davis, Albert Roy, and Rawls, Jr., Walter C. *Magnetism and Its Effects on the Living System.* Hicksville, New York: Exposition Press, 1974.

Day, Langston, and DeLaWarr, George. *Matter in the Making.* London: Vincent Stuart Ltd., 1966.

Dreiband, Susan, and Nicollela, Roberta Jeanne. 'Frances Farrelly on the Nature of Radionics.' *Psychic Observer* (Sept.-Oct.), 1975.

Eeman, L.E. *Cooperative Healing.* London: Muller, 1947.

'Expedition in the Sahara.' *Reader's Digest*, October 1958.

Finch, W.J. (Bill). *The Pendulum and Possession* (Revised edition). P.O. Box 1529, Sedona, Arizona: Esoteric Publications, 1975.

DeFrance, Henry. *The Elements of Dowsing.* London: G. Bell and Sons, 1948.

Friedman, Isidore. *The Mathematics of Consciousness.* Brooklyn, New York: The Society for the Study of the Natural Order, 1974.

Friedman, Isidore. *Organics: The Law of the Breathing Spiral.* Brooklyn, New York: The Society for the Study of the Natural Order, 1974.

Friedman, Isidore. *The Practical Use of the Pendulum* (a taped lecture). Brooklyn, New York: The Society for the Study of the Natural Order, 1976.

Forlong, J.G.R.: *Rivers of Life.* London: Bernard Quaritch, 1883.

Gardner, Martin. *Fads and Fallacies.* New York: Dover Publications, 1952.

Graves, Tom (ed.). *Dowsing and Archaeology.* Turnstone Press, 1980.

Hall, Manley P. *The Secret Teachings of All Ages.* Los Angeles: The Philosophical Research Society, 1962.

Hills, Christopher. *Nuclear Evolution.* London: Centre Community Publications, 1968.

Hills, Christopher (ed.). *The Supersensitive Life of Man. Dimensions of Electro-Vibratory Phenomena.* Vol. 1. Victor R. Beasely, Ph.D.; *Energy, Matter and Form.* Vol. 2. Alastair Bearne; *Supersensonics.* Vol. 3. Christopher Hills, 1976.

Jacob's Rod, 1693 (translated from French). Reprint. Boston: A.H. Roffe, Co., 1887.

Jaegers, Beverly C. *The Extra Sensitive Pendulum.* St Louis: Lumen Press, 1972.

Korzybski, Count Alfred. *Science and Sanity.* Lakeville, Conn.: International Non-Aristotelian Library, 1933.

Landon, Guy. 'The Nature of Dowsing.' *Psychic Observer* (Sept.-Oct.), 1975.

Lethbridge, T.C. *The Power of the Pendulum.* London: Routledge, Kegan and Paul; Boston: Henley, 1976.

Long, Max Freedom. *Psychometric Analysis.* Vista, CA: Huna Research Publications, 1959.

Long, Max Freedom. *The Secret Science Behind Miracles.* Vista, CA: Long, Huna Research Publications, 1948.

Lakhovsky, George. *The Secret of Life.* Mokelumne Hill, CA: Health Research, 1974.

MacDonald, Howard Breton. *The Pendulum Speaks.* Ontario: Provoker Press, 1969.

Maby, J.C., and *The Physics of the Divining Rod.* London: Bell, 1939.

Mager, H. *Water Diviners and Their Methods.* London: Bell, 1931.

Manas, John H. 'The Rod and the Pendulum.' *Psychic Observer,* January, 1971.

Mannhardt, W. *Die Götter der Deutscher und Nordischen Volker.* Berlin, 1860.

Matacia, Louis J. 'Dowsing Introduced to the U.S. Forces.' *Psychic Observer,* January, 1971.

Mathieson, Volney. *Super Visualization.* Privately published. New York, 1960.

Mermet, Abbé. *Principles and Practices of Radiesthesia.* London: Watkins Publishing, 1975.

Ostrander, Sheila, and Schroeder, Lynn. *Psychic Discoveries Behind the Iron Curtain.* New York: Bantam Books, 1971.

Pendulum (21 issues) London: Markham House Press, 1960s.

Richet, Charles. *Thirty Years of Psychical Research.* New York: 1923.

Richet, Charles. *Our Sixth Sense.* London: Rider and Co., 1930.

Richards, Guyon. *The Chain of Life.* London: Leslie J. Speight Ltd., 1974.

Russell, Edward W. *Report on Radionics*. London: Neville Spearman Ltd., 1973.

Strutt, Malcolm. *The Theory and Practice of Using the Pendulum*. London: Centre Community Publications, 1971.

Tansley, David V. *Radionics and the Subtle Anatomy of Man*. Health Science Press, 1972.

Thompson, Clive. *Site and Survey Dowsing*. Turnstone Press, 1980.

Tompkins, Peter, and Bird, Christopher. *The Secret Life of Plants*. New York: Avon, 1974.

Tomlinson, H. *The Divination of Disease*. Health Science Press, 1958.

Tomlinson, H. *Medical Divination*. Health Science Press, 1966. Health Science Press, 1966.

Trinder, W.H. *Dowsing*. London, 1939.

Tromp, S.W. *Psychical Physics*. New York: Elsevier Publishing Co., 1949.

Vallemont, Abbéde. *La Physique Occute*. Paris, 1693.

Vitvan, (Ralph M. DeBitt). *The Natural Order Process* (3 vols.) Baker, Nevada: The School of the Natural Order, 1968.

Westlake, Aubrey. *The Pattern of Health*. Berkeley and London: Shambala, 1973.

Wethered, Verne D. *An Introduction to Medical Radiesthesia and Radionics*. London: C.W. Daniel Co., 1974.

Wethered, Verne D. *The Practice of Medical Radiesthesia*. London: Fowler and Co., Ltd., 1967.

Wethered, Verne D. *A Radiesthetic Approach to Health and Homoeopathy*. London: York House, 1961.

Willey, Raymond C. *Modern Dowsing*. Esoteric Publications, 1976.